THE WINTER WOODS

Pileated woodpecker

THE WINTER WOODS

John R. Quinn

Illustrations by the Author

THE CHATHAM PRESS, OLD GREENWICH, CONNECTICUT

DEDICATION

To Pat, who, in the best tradition of the author's wife, did indeed type this manuscript for me, but who also survived eight New Hampshire winters with me as well.

ACKNOWLEDGMENTS

The author would like to thank the following persons and organizations for assistance in the preparation of this work: the staff on the Squam Lakes Science Center, especially Director Robert Nichols and Naturalist Earl Hanson for contributed material; neighbors Lew and Dennis Downing, who as loggers and conscientious hunters, gave me a broader perspective of the winter forests; Dr. Alastair Craig for his frostbite expertise; my editor Chris Harris for his astute criticism of the first draft; and last, my three children, for their unflagging, if shivering, interest in the January trail.

Excerpted material appearing on the following pages is hereby gratefully acknowledged:

Page 11: from *Snowshoe Country* by Florence Page Jaques, copyright 1944; The University of Minnesota Press, Minneapolis.

Page 33: from *"Ain't No Crisis Here!"* by Jesse C. Gill, reprinted from the *New Hampshire Times*, Concord.

Page 60: from *Selected Poems* by Leonard Cohen, copyright in all countries of the International Copyright Union. All rights reserved. Reprinted by permission of The Viking Press, New York.

Page 63: from *John Burrough's America*, edited by Farida A. Wiley, copyright 1967, The Devin-Adair Company, Old Greenwich, Connecticut.

CONTENTS

INTRODUCTION

Take a walk in the January woods. It is a quiet world of greens, greys, and white, seemingly devoid of all life and activity. But is it? The wind gives the bare branches of towering oak and elm a life of their own, a motion and sound known well in Northern regions. Winter — a time of rest for most of nature's wild things, a time of struggle and hardship for others when a blanket of snow conceals most evidence of conflict and survival in the winter woods, but not all.

The snowshoe and the cross-country ski provide access and a passport to this fascinating realm of activity. To learn to read the visible signs of winter life is to acquire an awareness of nature at her harshest and, at the same time, her gentlest, for many of the plants and animals most evident during the lush months of summer enter a period of necessary rest during the dark winter season.

A traveler moving through the still forest of January will find himself in the place of the Indian who had tenancy here long ago. The early peoples were dependent upon the most subtle manifestations of plant and animal life to assure their own survival. The modern hiker is not likely to have such a practical need for a sensitivity to nature at this forbiding time of year though contact with the continuity of life in the snow and the cold will greatly enhance any woodland venture.

The forested area of North America, in spite of civilization's inroads, is still vast in scope and variation. Some northern states such as New Hampshire, are seventy to eighty percent forested today, and here, as in Canada, one may quite easily become lost, far from the sights and sounds of civilization. But one does not need to travel the distant wilderness in order to experience winter's touch in the forest. Second- or third-growth wooded plots in suburban Connecticut, New Jersey or Pennsylvania will yield, for the interested observer, a wealth of information and experience; each supporting a different array of life depending upon latitude and the severity of the climate. Some of my fondest memories are of boyhood excursions into mere scraps of woodland not more than five miles from New York City. Indeed, winter or summer, the easy accessibility to forested areas, even today, allows their exploration and study to be rewarding for the amateur naturalist with strictly limited time and mobility.

Within the boundaries of 'snow country', the woodlands take on a number of varying characteristics. Those in the more Northern latitudes are called boreal forests and consist largely of coniferous trees such as hemlock, fir, tamarack and spruce. As one travels south, this habitat gives way to a mixture of conifers and deciduous hardwood, including birch, oak and maple. This is the nature of the forests in my own area of central New Hampshire. As one progresses further down into the regions feeling a lighter touch of winter, hardwoods predominate, here being called the hardwood or deciduous forest; most woodlands of Southern New England, New Jersey, Pennsylvania and Ohio are of this type.

For practical purposes, our discussion of the winter woods will focus on those areas more or less covered by snow and affected by sub-freezing temperatures during the whole course of the season. The wildlife and plant species I mention are common in an area bounded by the Great Lakes to the west, southern Canada and Newfoundland to the north, and parts of New Jersey, Pennsylvania and Ohio to the south. Many of the same or similar species are also found much further west as well. The specific observations in this book have, however, been made in a particular area, roughly fifty acres in extent, in central New Hampshire though it is interchangeable with terrain found all over the northeast. Here, in a spot well known to me, where there are beaver ponds and deer yards, rock ledges and wandering old roads, abundant food and protective cover is available for the native wildlife in winter. Here in the 'fifty acre woods', at a latitude where winter makes itself known sternly, shall we observe how living things adapt to the reality of a cold and spartan season.

So venture into the woods in winter. Travel with keen awareness in your ears, your eyes, your heart. Life is not gone here, but much is asleep or hidden, awaiting the return of the sun and warmth, and signs of its presence are subtle and obscure but as rewarding in winter as in summer to those who would seek them out.

EARLY WINTER
A Time of Preparation

*"The flakes are almost like spring petals falling.
This afternoon has a feeling of promise to it, such
as April gives us but it is a far more mysterious
promise than that of spring."*
— *FLORENCE PAGE JAQUES*
Snow-shoe Country

It is snowing. During the night the storm came; a precipitation fine and granular dusting the crisp leaves of November with the thin, sugary powder of winter's first whiteness. The ducks have left the river and are gone, moving to the coast fifty miles away in search of open water. From thence they will move south to the more gentle climes of the Carolinas and the bays and estuaries of Georgia.

It is much too early for the encumbrance of snowshoes, so I go on foot. Beyond the small pond where, sixty years ago, ice was cut for the use of the Boston and Maine Railroad, lies the forest. Though cut over a while back, judging from the three-foot-wide stumps still evident, the woods remain — now stoic and grey. The riotous colors of autumn are past. The season of red-capped hunters and the sharp reports of shotgun and rifle is gone; the woods are silent, waiting for winter as they have for thousands upon thousands of years. Now, winter is here and the invisible whisper of snow through the lattice of branches has announced its arrival.

Northern barred owl

Logging roads fan out through the trees, providing ready access to fifty acres of mountainous terrain, an infinitesimal part of the whole, but a tract I have come to know and respect. I have been often to its wooded recesses, in May when the glades ring with the flute music of the wood thrush, during the green-gold days of summer, the air heavy with the scent of life and sprinkled with the star-bright bodies of flying insects caught in the afternoon sun.

Where does all this rich life go when winter's forbidding touch arrives? That eternal question endows the winter woods with a special challenge to me, for now I must focus my senses to a degree not needed during those days when the creatures of the woods assault consciousness with their color, movement and sound. Walking here now, under leaden sky and driving snow, I feel the grim yet natural and efficient preparations of nature. Most leaves have long since returned to the earth with only the beech providing subtle color against this grey place. The dried, twisted leaves, a pale buff, will remain on the trees through the coming season. The wind will free some and send them racing across the snow in February, but most will persevere until the swelling buds of spring displace them.

All deciduous trees in the temperate woodlands must be dormant by mid-winter or they will freeze and die. As a visible part of that process, the annual leaf change is perhaps the most strikingly beautiful of all the winter preparations in the natural world. Responding to the lengthening nights and steadily lowering temperatures, a layer of relatively insensitive tissue is formed between the leaf stem and the parent branch. With this interception of the water and nutrient supply, the leaf's chlorophyll is not renewed and it loses its normal green color, exposing the other, often brilliant pigmentation present within its flesh. When cells have weakened sufficiently, an errant breeze breaks the brittle union and the leaf spins down. Though evergreen trees do not shed all their leaves at once, a periodic process of 'rearrangement' goes on. As it grows, the tree abandons its lower branches, shedding the needles first, so as to discard those leaves too shaded to perform the act of photosynthesis.

Crossing a small park as I start out, with nothing but a low, pale sun in the east as company, I see the old ice pond beyond, black and brooding, shaped roughly like a giant hourglass, its quiet war with the ice just beginning. The pond's level is much lower than at the turn of the century when men with great saws took blocks of the good, clear ice, loaded them on the horse-drawn sledges and carted them off for the benefit of the Pullman passengers. Now it is but a mere bulge in the brook that feeds and drains it. Trout live here still and in the shallows, where the feather-delicate new ice creeps out, the water's surface is dimpled as dace nose curiously at the falling flakes. In the darkness below, where the cold has already taken hold, fish, tadpoles and insect nymphs of all varieties burrow into the protecting leaf litter and mud, there to await the sun's returning warmth.

The goshawk is a bird of speed and power, equal in most ways to the storied peregrine falcon. Two days ago, one of these fleet creatures, true "hen hawks", flew across our garden bearing a chunky but unidentified bird in its talons. Today, leaving the pond and skirting the park which, it seems, only recently rang with the cries of scampering children and beer-swigging youths, I approach a great maple stump surrounded, as a knight by vassals, with stunted alder shoots. These are festooned heavily with white downey feathers, as though someone had killed a white leghorn there, scattering its plumage carelessly to the wind. The truth is not far removed, for connecting the two occurrences, I quickly realize that here was the destination of that grey hunter, and examination of some of the body, or contour feathers of the victim, prove it to have been a rock dove or street pigeon, one of those camp followers of man that find shelter and safety near his works. This short-winged, long-tailed hawk, driven by hunger, had invaded the town and had snatched the pigeon, perhaps right off a bird feeder, and bore it off to the edge of the forest to pluck and devour it. Not much remains here to tell the story save some gore-clotted skin and feathers and those little downy flags fluttering in the chill wind from every twig and stem. When these winged hunters of the deep woods venture so close to us for prey, one can expect a hard winter.

With a curious sense of expectancy, my path wends it way out of

The winter woods

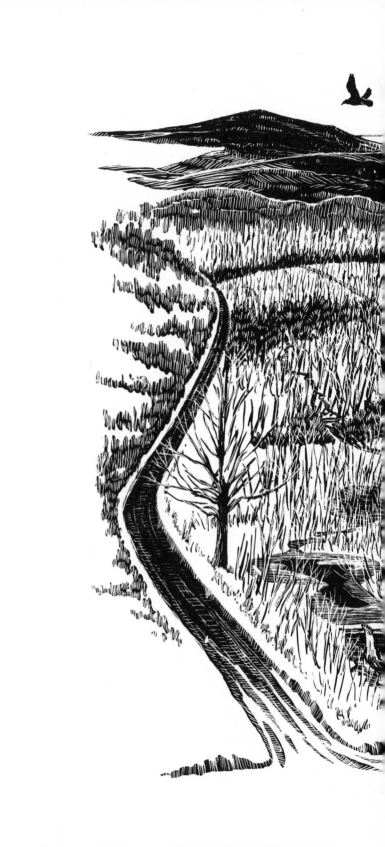

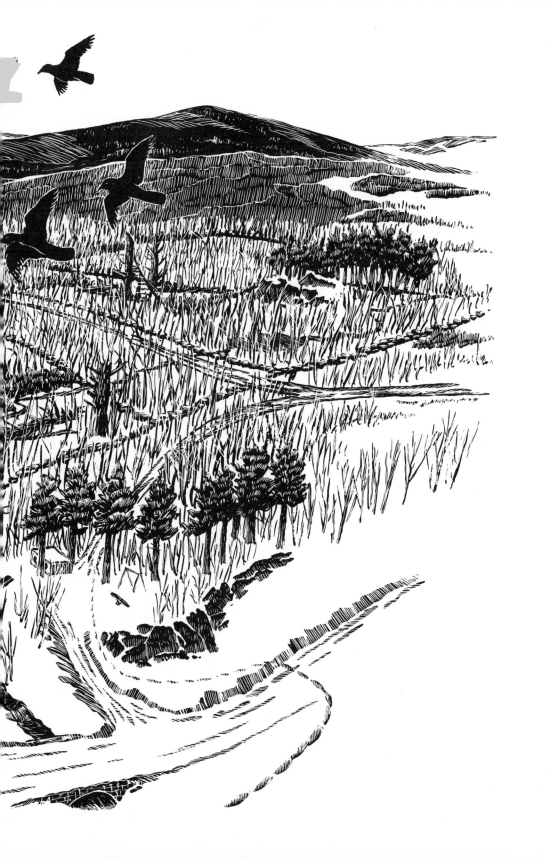

the park and into the woods. The term *woods* is the only one that can be employed here with any accuracy. The word *forest* will just not do with its conjured visions of virgin stately giants of pine and elm, untouched by saw or axe. Here, land has been cleared and cut, and the resulting second growth is pathetically spindly, with most trees having less than ten inches at their base. In the snow-bearing wind their bare branches rattle and sway, squeaking and groaning against each other; the music of the winter woods. With the increasing cold, the life- giving sap withdraws down toward the roots, the source of their life, leaving the outer branches to grow stiff and cold so that on a frigid starlit February night, one might snap off a living twig as easily as he might one long dried and dead. In the hundred-year-old logging road I follow up through this glade, my imagination wanders back to the time when these woods were a source of timber and income. What hardy breed of men drove oxen and grey horses over this path? The ruts are still here, leaf-filled and nearly obliterated, but evident as monuments to the works and dreams of those long dead.

The leaf-litter, not yet fully covered by the new snow, is noisy, announcing my approach far and wide to all those creatures within hearing. The leaves lie in loose, dry, windrows that will soon flatten under the weight of the snows, but for now they whisper and rustle as they did through October, giving voice to the presence of man or beast. A lone fox-sparrow scratches industriously in the litter ahead. Most of its kind have long since passed to the south, leaving this fellow to share the forest floor with the hardier juncoes which will brave this place for weeks afterward until they, too, leave for less hostile lands and more plentiful food.

The brook that traverses this woodland is still active. At the place where the logging road crosses it, the water spreads out over a silt bottom and moves with a placid motion, but above and below the crossing, it stumbles and rolls over rocks covered with fontinalis, that cold-loving aquatic plant of the woods. This intensely green plant, found in clear cold mountain brooks as far south as New Jersey and Pennsylvania, adds a touch of color to early winter's drab vista that is otherwise unrelieved except for the more subtle green of pine and hemlock and the tans of beech and oak. The brook will flow open to the sky until cold and snow hide its activity under a mask of white, but even then, occasional openings down through the snow will provide drinking places for those creatures remaining mobile through the long night of winter.

Just to the south, where this brook joins the other that enters the pond, I come upon a low spot, a place where the water spreads out over an area of rushes and alder thickets. Here sculpins, small, big-headed fish with great fan-shaped pectoral fins adapted for life amid stones and weeds, remain active, probing for insects and small larvae. Woodcock frequent this bog in spring and summer, searching for earthworms in the wet soil and rearing their broods in the surrounding brush. Now the partridge, their summer done, move into the area and prepare for hard times in a place that will provide them with food and shelter long into the winter. Stopping, my eye is caught by the very slightest of movements near where the softly flowing stream covers grasses, yellow and brown, bent by the flood and submerged in black water. It is a splendid cock partridge. The bird walks stiffly erect, his crest high and his slightly fanned tail jerking with every step. He has seen me and awaits my next move, as if to confirm the danger. So perfectly do his colors blend with his surroundings that he is but an extension of leaf and log, set free and moving through the snow and grass. Try though I may to maintain an absolute stillness and silence, some small motion, unknown to me, expresses a threat and the bird explodes into flight. Banking steeply, dodging through the saplings, the bird streaks off with the powerful flapping-sailing flight peculiar to the species, and disappears like a ghost into the gathering storm.

Today one can almost see the exact beginning of the new season all about in this sober landscape. The partridge move with an activity born of urgency. The accessible ground-cover still yields a harvest of bright red partridge berries, a favorite of these ground-loving birds, and in the bog itself, the last, lingering insects are sought out and eaten.

The creatures of the brook are growing torpid and sluggish, seeking the sheltering mud and debris of the more quiet, deeper spots in-which to spend the winter. They will enter a state of true dormancy, their respiratory rate determined by the surrounding temperature. The tiny, black-striped dace, which give the summer brook such vitality and action, will vanish over a period of a few days as water temperatures hover near the freezing mark. Breathing stops, or nearly so, and the fish enter a state of suspended animation tucked away under leaves and silt, surviving even short periods of ice provided that their life-giving vascular systems and tissues are not frozen. One has only to stick a finger into the frigid water of a winter brook to marvel at the tenacity of life displayed by such small scraps of flesh and bone. We may retreat to the comfort of warm kitchen and crackling fire while the fish are where they are for the duration!

In the growing light I follow the trail until it bisects an old stone wall, a seemingly useless work of man here among the trees. But sixty years ago this was fertile field, cropped and controlled by the appetite of sheep or cow, or perhaps filled with the rustle of corn in summer. As time and tax rates changed, the land was abandoned, or sold, and allowed to return to its original crop, the trees. Slowly they came back, spindly and vulnerable at first, then great and strong, then harvested too as produce. The logging road I follow offers testimony to that fact and here the enduring boundaries of piled granite remain, monuments to the energy and struggle required to prepare this hostile New England landscape for the plow. The walls provide home and shelter for that ubiquitous little sprite of the woods, the chipmunk.

Chipmunks don't hibernate. At least this is a recently advanced theory, refuting the accepted belief that the rambunctious little ground squirrel is one of New England's seven sleepers (or hibernators), among the mammals. The others are meadow and woodland jumping mice, three species of bats, and the woodchuck or groundhog. The chippie was long believed (and still is, depending on the authority you read) to enter a true state of hibernation deep in its subterranean burrows beneath tangled tree roots and stone walls. But recent studies indicate that although these animals, like raccoons and skunks, will sleep for weeks at a time through the severest weather, they then awaken for short periods to feed on the nuts and buds in their underground hoards. But rarely, if ever, do they emerge above ground during this inhospitable season, which is why you won't see a chipmunk trilling from a snow-covered stump in February.

Today though, a flash of russet catches my eye. Along the stone wall, with its thickening skin of snow, a tardy individual, cheeks bulging with last minute delicacies, hurries with rapid jerking pace. Aware of his conspicuous image against the whiteness, he senses his vulnerability as I stand not five feet from him. The grey form of the goshawk might flash through the trees, unseen until too late, and pluck him off the rocks. The sharp eye of a hunting weasel, a predator active all year, could pick him out easily now. Snow is not a time for chipmunks. He must reach the cleft in the wall and the entrance to his burrow and make this his final trip. This time death does not come out of the grey trees. Seeing me at least, the chipmunk gives a chirp of alarm and vanishes underground. For him too, winter is here at last.

The few ill-tempered woodchucks that live at the edge of the woods, near the broad field and the road that climbs to the mountain, have since denned up and will not be seen again until the snows retreat and green plants venture forth. The chuck is a true hibernator. This lumpy but powerful rodent with grizzled brown pelage enters the complicated deep sleep we wrongly assume belongs to the bear. Depending upon the length of the winter season and the depths to which temperatures plunge, the hibernating chuck's body temperature will drop to forty degrees or less, and respiration may occur only once every ten to fifteen minutes. No alarm clock in the world will awaken this slumberer. The accumulation of its insulating and energy-giving fat as well as declining temperatures, trigger the hibernating instinct and nothing less than a reversal of these factors will bring the sleeper back to consciousness and activity. September and early October are periods of non-stop eating for woodchucks, and it is especially evident in those that inhabit more northerly areas. They must build that critical layer of fat which, as in the case of all other hibernators, is the chuck's primary defense against the cold and the absence of available food.

The poor, persecuted woodchuck! In farming country, where

they are most prevalent, yet most unwelcome, they lead lives fraught with constant danger and are worthy adversaries of the .22 rifle indeed. I recall my own single such experience with one particular chuck and his untimely end at the hand of man.

Accompanying a farmer friend on a woodchuck hunt in the spring of last year, I was confronted with a nearly perfect example of varying, yet side-by-side attitudes toward the land and its creatures. My friend hunts the chucks in his pasture for one reason; they regularly raid his gardens and he regards them as a threat to his well-being and would thus be glad to exterminate them totally. My own view had been that while I fully realized the havoc uncontrolled woodchuck hordes can wreak upon unguarded produce, the flabby creatures have such endearing qualities (anthropomorphism at its best I'm sure) and the thought of my killing one so unappealing, that I probably would allow them to munch away contentedly in my own garden or, at best, live-trap them and spirit them elsewhere.

But I felt a traitorous and undeniable excitement as we stalked the first chuck, which I myself picked out against the rolling sward 100 yards away. A sense of guilt and regret overcame me, as the animal, sitting rigidly upright at the mouth of his burrow and thinking himself quite safe so close to home, watched us approach with blissful ignorance. At fifty yards, my friend assumed a sitting position, cradled the rifle (I carried none) and squeezed off a shot, which struck the woodchuck in the head. The animal dropped out of sight and all was quiet. It was a clean kill; the slug penetrated the braincase and death was instantaneous; the groundhog had no idea what hit it.

Having no real use for it, my companion gave the carcass to me, and utilizing my moderate skill in preparing museum study skins, I dressed it and we enjoyed our first woodchuck stew. It was quite good and though I've eaten game birds and animals before, for some vague reason, this particular incident altered my view of the death of an individual wild creature more than any other I'd witnessed. Perhaps in growing older, I began to look at nature less from an emotional or idealistic viewpoint and more from a realistic one. Death is everywhere in nature; it surrounds her wild creatures (though they know nothing

of it), and its final touch may be imminent at any given moment. It may come from without, as in the swift strike of a predator, or from within, as the touch of disease, hunger or the activities of internal parasites take their toll. The conflict I felt is contained in the question as to whether modern man, with his superior intellect and increasingly technological advantage over his quarry, has evolved himself out of the moral right to participate in the hunt as the lesser creatures pursue it. A naturalist such as I claim to be, well schooled in the aspects and idiosyncracies of wild places, must sometimes struggle to retain a sense of man's true place in the scheme of things. We must inevitably, to accomplish our own survival, take from the land, but in so doing we must hold it in respect and try to operate within the systems of checks and balances as wild creatures do instinctively. As our world-wide crises of pollution, and food and space shortages vividly show, nature offers only harsh reprisals for the abuse of her bounties.

The early hours of the day are slipping past and the snowfall has begun to thin. The pallid sun, though still low on the rugged horizon to the south, is nearer its zenith and sheds its light now through a sky of frosted glass, giving the scene a rainbowed essence of something primordial and unworldly. Every flying snow crystal, indeed the very air itself, seems shot with bright color and charged with a sense of expectancy. As if in answer to this bit of forest magic, the silence is broken by a sound one could associate with nothing other than the rich tympany of sleigh bells. The noise grows, fades and then is all about me as a rolicking group of chunky, piebald birds pass overhead, just above and through the treetops. They keep up a constant rolling chatter and their black, gold and white bodies, which appear entirely too tropical in their bright pattern to be found in such a place, lend a touch of color and warmth. These are evening grosbeaks, birds of winter that eagerly visit feeders stocked with sunflower seeds and are happiest in the sheltering greenery of conifers. The appearance of another small group of more somberly-hued birds, pine grosbeaks, promises that this will be a 'finch winter', their early arrival foretelling deep snows this year. These big, grey and rose grosbeaks linger in the topmost branches of an ancient apple tree here at the wall, stripping buds and working industriously to open holes in the hard skins of the few frozen apples remaining. One bird leaves, launching itself into the air and flying off with a strong direct flight. Another follows and at regular intervals the members of the flock abandon the apple and move off in a line, like a string of Christmas lights yanked off a drying tree in January.

Without the birds, the winter woods would be bereft of most of its sound and motion, for the only other group of active vertebrates, the mammals, are much more secretive and most of these are active at night, no matter what the time of year. But even on the coldest, greyest day of the year, one can manage to scare up a few hardy chickadees or a white-breasted nuthatch, hitching its way up and down the tree trunks in search of dormant insects and their larvae. It has been said that if chickadees weighed a pound, they could take over the world. Standing in the woods on a below-zero day, surrounded by an agitated gang of 'dees' that you've squeaked out of the surrounding timber, you can easily believe it!

Resuming my walk I encounter an ancient 'blowdown', a tangle of great pines and elms that have fallen to some forgotten September gale. In death they provide life in many ways. The food and shelter found among their exposed, snarled root systems and within the prostrate trunks is critical to many animals in winter. Mink, weasel and even bobcat will find ready refuge here; safety from the talons of great horned owls and the guns of men. These giants, three feet in diameter, have been down for many years and probably weakened long before that by age and disease. As great slabs of bark fall away like the skin of a flensed whale, the mosses move in, colonizing the tree from windborne spores and marching up from the forest floor. Gripping the soft wood, their roots penetrate and assimilate the cellulose and begin its return to earth. The great tangled root systems stand vertical now, clutching rocks of all sizes tightly in their dead fingers. Soil has long since been washed away to be replaced by the ubiquitous club mosses and bracken fern. Looking closer here I see my first tracks of the season. Flung through the snow like a necklace of some delicate design are the tiny prints of a short-tailed shrew. They expose a meandering, but somehow intense purpose; that of a tiny hunter in search of a victim. The tracks, for all the world like those of some miniature windup toy, hug the rocks and recesses of the upturned roots and then vanish into their tangled depths.

Shrews, which strongly resemble mice in form and action, are insectivores rather than rodents. They remain active throughout the winter, dependent on a set of dentures fully capable of dispatching prey as large as they are, and subject to a very high metabolic rate which makes them veritable dynamos of energy on an endless search for food. During the summer months these savage little sprites dine on almost anything they can overpower; insects, spiders, young mice and snails are common fare. It has been estimated that if an average human were to eat as much as a shrew does in proportion to his size, he would have to put away between 100 and 200 pounds of food a day! The smaller a warm-blooded animal is, the faster it loses body heat through radiation so that, in the depth of winter, this problem is no small one to be overcome. Standing here, gazing at the tracks and intermittent tunnels of an animal that will, if it's lucky, see a second winter, I feel compelling admiration for such tenacity of life.

This brief, first snow is ending, tapering off to a thin granular sleet that whispers dryly through the thickets. Near the crest of a modest hill just ahead, capped by a rugged granite outcrop (the tip of some lost Devonian mountain?) there stands a dense cluster of hemlock. Such a grove is always a place of shadow and safety, an oasis, in a world fraught with the perils associated with open, exposed space. For all nature's creatures, the interlacing shelter of evergreen provides both respite from heavy snows and a place of hiding and rest; for owls and other animals given to nocturnal movement, the green seclusion affords relief from the ubiquitous chickadees and bluejays that seem to delight in depriving the night hunters of their diurnal slumber.

One approaches a thicket with a sense of expectation. Coming upon it, in the midst of a relatively open forest, the thought occurs that there *must* be something secretly hidden here, something that will, at the last minute, burst forth with sound and jolt and flee on foot or wing leaving us in heart-stopping silence. Protection may be pro-

Short-tailed shrew

vided to smaller creatures, but a dense stand of evergreens will most likely not produce the startled Bambi or scuttling Thumper we might like to expect here. The woods are not filled with the little furry people as in our childhood books. But here, sheltered, the living things are indeed present.

On a branch of juniper is the hanging, silky sack of the bagworm, a solid larva that feeds on this shrub through the summer and fall, spinning itself a snug warm home in which to weather the cold. In spring it will emerge as an adult moth. Seldom straying from the tree of their birth, bagworms undergo a rather complicated life cycle, completing several molts. The shed skins are pushed through the open lower end of the bag as are all wastes and feces. The several species, each of which show a separate preference for either deciduous or evergreen trees, incorporate twigs and bits of leaves of the host tree in the bag-building so that the result often has a distinctly festive look.

The storm is past. No snow at all slants though the quiet trees and the sun pokes strong fingers of light in and out of fast moving cloud cover. The frail crystals begin to melt away; it is not yet time for the permanence of drift, and white ice. Lush branches of pine and spruce shiver and sigh against a clearing sky. Surely I'll discover something here, some furtive beast in refuge amidst the sun and shadow!

An owl. A slab of bark, rough and streaked. A dusky, almost supernatural extension of a tree. The bird, a barred owl, stirs and opens its bottomless black eyes, and gazes at me from its perch high in the bracken. I, the unwelcome disturbance, an unexpected interference in the rhythm of sleep and hunt, or so I feel. Or am I! The stolid bird, void of hostility, watches me with innocent curiosity (shall I walk around it and watch it turn its head an impossible 180 degrees?). I want to stand and watch it until my eyes ache or until the bird gives up and flies away, attracting pesty crows and jays in its passage. I've seen many a barred owl, but each time their wonderful stygian eyes, unblinking and soulful, fill me with a kind of religious kinship. So trusting and gentle (this big owl kills only mice), they move about and hunt in darkness or on days of grey overcast and display little real fear of people and are slow to learn of the danger of guns.

Somehow, owls make me think of winter. Superbly outfitted for a life of survival in the cold, grey lands, they pursue their bright-eyed, somber-hued lives in solitary intensity. Alert for any secretive motion, they hear and see things we civilized bipeds could not hope to. Fluffy, insulated, bulky feathering covers a muscled body only the size of a bantam hen with powerful legs which can kill with the sureness and skill of a practiced executioner.

The quiet bird, shifting nervously from foot to foot and fixing me with liquid gaze, takes me back to other winters, other places. Green hummocks of stunted pine against the dry browns of spartina and fox-tail grasses of Long Island's barrier beaches; snow and ice in hollows and dune buggy tracks, purple leatherleaf and heather whip in the sea breeze. Winter's touch seems lightly felt. Here I found another creature that hunts and flies in crepuscular hours, within the sound of roaring surf and windblown sand; the little, gnome-like saw-whet owl.

This rotund midget, so soft of wing and demeanor, is a skilled hunter of mice and an occasional song bird. At Cedar Beach, the owls winter in some concentration (for owls) among the low windblown clumps of pine that dot the grassy expanses just behind the dunes. They prey here upon the millions of mice that ply their whispery way along well marked by-ways through forests of phragmites and rushes, retiring at last to their evergreen hideouts.

The owl, sacred to the Greeks as a form of Athena, has long been considered a creature of mystery and super intelligence. The ancients believed that "any life issuing from crevices or holes in the rock would be her (Athena's) life," and that the goddess had special insight and affinity for all life on earth. A noble mission to be sure, but one that most owls seem unaware of as they pursue their nocturnal occupations. Owls do, however, display an adaptability and tenacity that must be admired. Most species remain where they are throughout the seasons, and fare quite well as long as the population of their prey species remain high enough to sustain them. Wading strenuously through knee-deep snow, in the silence and snapping cold of January, I have often been compelled to wonder how a large and active predator such as the great horned owl is able to find enough food to sustain its metabolic fires. The body of any bird is a highly efficient radiator of heat and without its plummage of warmth-holding down and contour feathers our owl, in addition to looking ridiculous, would quickly perish from the cold. Even in relatively mild weather it would lose more warmth than it could produce. In the extreme temperatures of mid-winter nearly all birds and mammals remain as inactive as possible in order to extend their precious reserves of energy and body fat. Undue disturbance by humans at this time is a prime danger to wild creatures with rationed resources.

My barred owl has had enough of me. In spite of my sincere effort to withdraw without disturbing him, he has misread my movement and, with a light swish of pinion and pine needle, leaves his perch and glides off through the trees. Examining the ground below the vacated perch, I see evidence that this is a favorite roost. Pellets litter the area. Lumpy, greyish, thumb-sized objects, owl pellets are a sure sign of an owl roost, winter or summer. For anyone with a strong stomach and scientific interest, the dissection of pellets collected under such a site will provide fascinating proof of the feeding habits of most owls. Delicate mouse bones and skulls along with the indigestible hair form the bulk of the average pellet, with a few bird beaks and toes thrown in, depending upon the species. And it should be mentioned that such items leave the owl in the same manner that they entered.

The day is waning and the lowering sun lays great, pale, gold ribbons across the new snow. I must leave the woods for now, but will take a circuitous path to prolong my day's journey as long as possible. Hunger has begun to make itself felt in my innards along with the thought of how easily it is satisfied for most of us. Thirst is not a real problem, as the woods are laced with brooklets which, here at least, are potable and have that distinctive sweetness of moss and pine, with the near-effervescence of water charged with oxygen, dark and cold.

Fox tracks! They are fresh, laid down but minutes ago with edges sharp and snow crystals dropping into each circular print. They stretch away in the direction I wish to go, showing casual yet purposeful intent. I can almost see their maker in my mind's eye, furtive yet proud, splendid brush flowing behind. A fox seems almost to flow over the ground, running with a controlled, fluid grace, a superb living mechanism finely attuned to even the slightest sight or sound in the surrounding environment. This fox, a big one from the size of the tracks, seems bent on exploring the old cellar hole that lies against a south-facing slope not far away. Once the haunt of men sometime in the middle of the last century, this boulder-lined hole in the ground, with its attendant deep-well, is all that remains of a thriving farm. Where once the root cellar held garden produce and preserved delights open to pilferage by the children of another age and time, now mouse, shrew and garter snake hold sway. And these, or at least the first two at this time of year, are what the fox will seek, deep within crannies packed with twigs and dried pine needles. Will a fox eat grapes as Aesop tells us? I think not, but then this wonderful little dog of the woods has been known to eat less appetizing fare, and most carnivores will eat vegetation when ill health requires it.

There! A flash of red and spume of snow powder! The splendid creature has sensed my approach and leaps from the old foundation. He almost flies, skimming the ground and disappearing into the jumbled chaos of a wind-fall. He will not stop here, I'm sure, but feint and dodge with skill, leave the fallen trees, use each concealing hillock and stump until he is far from me and all of the threat I embody. Farewell fox! It's just thirty-one degrees and winter is but beginning for you.

Another creature has been sharing the mouse-bounty of the cellar hole. With quick motion of blackish-brown, a long, low animal, measuring about a foot in length, scuttles off and away toward a nearby brook. A mink. Ferocious little hunter of mice and shrew, prized by fashion, he is without benefit of the white ermine color that the otherwise similar weasel assumes to camouflage him from hungry winter enemies. The mink will remain luxurious chestnut brown, which will certainly not render him invisible against the snow, though he will find enough safe cover in rocky, log-strewn stream beds and other hidden spots to see him safely through this perilous season. His head, like that of a furred, short-eared snake, peers at me over a snowy root, and is then jerked out of sight to be seen no more.

So the walk home continues, against a now-grey backdrop. The sun has gone, not behind the hill yet, but behind a gathering overcast. Snow is again in the air as it was at dawn. A fitting end to this early winter walk—a curtain of falling snow rung down upon the life of summer. In the higher branches of a thirty-year-old maple, there hangs a white faced hornets' nest, now very quiet and empty. It is acquiring a cap of snow and its high position in the tree portends (so the New Hampshire natives say) a severe winter.

The flat plain and abandoned seesaws of the park signify my reentry into civilization. Gathering darkness and snow have precluded any attempt to visit a "lost" beaver pond on this trip but perhaps another time. The rush of passing cars, the town with mailboxes, blue and white, on corners, hulking shapes of houses black with golden eyes, makes the forest once again seem foreign and far away, almost forbidding. Once we were an integral part of the woods, lived off and within them, and now we go as conquerors, or visitors seeking inspiration. Leaving the realm of the trees and the kingdom of winter, one can see just what we have lost, and what we can again regain.

By my back door, splitting wood for the stodgy 1884 Hub parlor stove that squats darkly in our living room, my wedge exposes a network of tunnels throughout the log's center. Packed here, in solid black ranks, are carpenter ants, in hibernation and sluggish. Big, active insects, with an impressive pair of biting jaws, these ants are not animals one would wish to introduce into the house, so rather than have any drop free and revive in the warmth inside, I gingerly scrape them out into the snow, there to perish overnight. We are not, after all, that far removed from the natural world. Surely it claims and influences us all.

Whitefaced hornet's nest

Mink

MID-WINTER

A Time of Hardship

On winter nights I set a spell,
My rocker creaks and groans;
Knotted wood snaps in the stove,
The northwind howls and moans.

My big sled dog sets by my side,
His head upon my knee;
The flicker of a candle's gleam,
Is all that I can see.

—JESSE L. GILL
From "Ain't No Crisis Here!"

The day began with bluejays. Garrulous and enthusiastic with feathers fluffed against the fifteen below zero morning, the jays were at the suet outside the house just after dawn, their strident voices cutting the crystal air, tiny breaths feathery plumes against it. A bird's exhalations are not of the greatest volume as one might imagine, but this morning they can be seen, lightly flashing upon the frigid air, as the hardy, adaptable birds squabble and jockey for positions at the swaying suet balls. Far off, across the tops of the unmoving trees, crow voices welcome what promises to be a real northern mid-winter day and a good one to try the woods on showshoes. Much snow has fallen over the past month. Winter has come in with a heavy hand, lending a ring of truth to the many rustic predictions of a tough winter that I heard spoken during the prior months; corn husks *were* thicker this summer; the white faced hornet nests I saw *were* located in the upper branches of trees; chipmunks seemed to carry their tails higher than ever (a sure sign) and our cat not only sat with her tail toward the fire, (an old Ozark belief), but she also spent a lot of time sleeping under the stove! And wooly bear caterpillars tended much more towards wide black bands this year. So we've hunkered down for the worst nature has to offer; no damp chill and sleet here, this is winter country!

Prepare for it! Three feet of snowfall, impassable roadways, unpredictable weather trends (Ben Franklin's Almanac notwithstanding), these things are all common aspects of a northern winter.

Before setting forth into the woods, there are certain winter rituals to be performed here at a latitude where the cold is a definite physical presence to be reckoned with. A New Hampshire transplant, it took me awhile to adapt to the necessity of certain survival techniques. Coming originally from New Jersey where snow season is often characterized by a lack of the white stuff, cold, damp and a decidedly less than pure air, and a definite dependence upon technology to keep one warm, it took me some time to adapt to the Yankee approach to "getting by" without selling one's soul to the oil and power companies. Here that is something just short of an art. Here in the north country, where winter extends from November to early April, wood supplements oil, and freezing water pipes, collapsing roofs from the weight of snow, and serviceable snow tires are things to be expected and dealt with. Ice forces its way up underneath roof-edge shingles to spread unsightly stains against ceiling tiles, and on the roads, salt is used with luxurious abandon, eating away fenders and rocker panels of northern New England cars. A biologist friend once ran a test on the spring gutter runoff and found it stronger in salinity than the ocean itself.

Wood must be split and stacked by the stove, else the crouching mechanical monster in the cellar will sing all day with its flaming tongue of oil and run the fuel bill up to astronomical proportions. During the coldest days we've had to allow the faucets to run at a slow dribble to prevent the freezing up of standing water in the pipes in the house's earthen cellar floor, a menace whose only other solutions are electrical tape wound around the endangered pipes, or another wood stove down below, with the attendant extra three cords of dry wood to keep it hot and useful.

Once the house has shaken off the night's chill, one can think of going into the woods. "You're going out today?" marvels my wife. "It's at least ten below out there and there's a wind, too!" Mumbling something about loggers doing it and the joys of cold weather hiking, I head for the barn where my "long shoes" hang in repose from a wall spike. Being a 220-pounder, I had long ago found that bear-paws, those smaller oval shaped snowshoes, just would not hold me up in deep powder, leaving me floundering helplessly. My own shoes, of the "Maine" brand, are five feet long with a tapered tail that leaves a distinctive line in the trail. They provide nearly the same stability as cross-country skis when one finds a steep downhill ledge to negotiate, though in the hands (or on the feet) of a novice, they can prove cumbersome when maneuvering over those fences or through wind-falls that regularly send the unwary sprawling facedown in two feet of powder snow. Regaining one's feet after such a fall can be an awkward procedure (carrying ski poles or at least a walking stick is a prudent forethought here), but if you are touring alone you can grunt and swear without inhibition.

Today I have not waxed the wood and webbing of the shoes, or even applied one of the "no-stick" sprays that are often used on snow shovels and the like. Everything today is so dry; each snowflake a thing to its own, merely touching, not joining its fellows. After securing the rawhide bindings I set out into the snow-bright morning, under a sky of poetically beautiful blue, studded with the rich green spires of pine and hemlock.

Limbering up, a swing through our own two acres across the road from the forest seems like a good idea. Three years ago, when clearing small trees (ignominiously called brush here) from a pathway to the brook, I decided to set up and maintain two brush piles for the use and shelter of lesser creatures that inhabit virtually any piece of land un-touched by lawnmower or plow. Here the piles sit today, huddled under snow and looking as snug as some mammoth muskrat lodge might be. But are they being used? I wish that I might somehow jack up the whole jumbled mass so that I might see just what and how many of the creatures of summer slumbered here or scampered through the intricate woody maze. In July I had released a small king snake here, after caring for it for a month; put a young white-footed mouse on a branch and watched him hop off into the dark pile to make his own way in the world. Have they survived? Are they down there now, below the cold, down in the leaves and roots and dark tunnels, the land of the winter sleepers? If not they, then others must be, for man-made brush piles, an unfortunate rarity in today's world of neat, antiseptic gardens and lawns, provide a welcome and necessary haven for small wildlife. Even a pile of 'slash cutting' in the deep woods, piles of branches left from logging or land clearing, will attract gatherings of mice, shrews, small birds and reptiles in much greater number than will the surrounding habitat. Though perhaps a bit unsightly, especially in more suburban surroundings, the brush pile makes good conservation-sense and will reward its builder many-fold.

Aquatic brush piles will prove equally valuable, attracting small food fish and invertebrates that in turn will bring in the game species sought by sportsmen. All species of catfish (bullheads or horned pouts, depending on where you are), bass and yellow perch will utilize the food and shelter provided here, winter and summer; the fallen tree in the moving stream, the clot of debris lodged in the turn of a brook, these are the places in which to look for the fish.

The snow has arrived now by the drift, white and soft, as we've yet had no thaw that will give it a crust of any kind to make snow-shoeing easy. No, this is nearly two feet of powder, snow that even with the supporting aid of the shoes will prove difficult to negotiate. Snowshoes seem a bit more like snow shovels in powder but no matter, the white stuff does have its redeeming qualities and is, in fact, absolutely necessary to the survival of plants and animals in northern places.

Trying to maintain some kind of rhythmic step in this fluff, with the stinging breeze reddening my face and singing past my ears, it's hard to imagine anything surviving a New Hampshire mid-winter, but many species of plants actually need a period of rest and dormancy in order to reopen their buds in spring. Below forty degrees, plants enter the state of dormancy simply because, at these colder temperatures, soil waters are unable to dissolve and supply them the nutrients needed for growth. The buds of a winter maple are protected by small waxy leaves that shield the infant leaves against drying (desiccation) which is actually a greater danger to the tree than the cold itself.

But it is the low growing plants and mosses beneath the snow cover in early winter that have things the easiest. They use the principal of insulation, and they use it in pretty much the same way as an animal would. A few inches beneath the snow's surface it is warm. Well, not *really* warm, but on a below-zero day with a stiff Nor'east wind on the move at ground level, temperatures under all that white will remain at thirty- to thirty-two degrees, day and night throughout the long winter. Thus the tender growing things of summer gain shelter below the snow's protecting mantle from the biting winds and severe cold that would otherwise quickly freeze and dry them.

Anyone who has ever snowshoed or skied a mountain top of even modest elevation will have seen, probably without realizing it, the Krummholz forest. This natural phenomenon, named for the German biologist who described it, is a perfect example of the insulating, life preserving qualities of snow. The trees of the Krummholz forest are of the same species as those further down the slope, but here, where winter is a demonic force, the trees adapt to the snow depth. A birch or mountain ash will start life as a seedling and grow through the seasons in the normal way until it reaches the level of the maximum snow drift. Then a change takes place. Winter winds and cold prune off any adventurous twigs that attempt to progress above the snow, and turn back other growth, so that a seventy-five-year-old balsam or spruce may be but two feet tall and a mass of twisted dense branches: the Krummholz forest, a thing of windswept places above the treeline.

The cold of winter is enough to drive almost any sign of growth and activity underground, and the mosses of rock and stream bed are no exception. Their life story, their coping with winter, is more an act of simply stopping growth and remaining static, rather than retreating or casting off leaves. The mosses are among the first plants to explore and colonize new and inhospitable home sites, so they must be able to meet the worst of conditions and thrive, crouching down under the weight of snow and emerging green and bright in April.

Swinging along, I think of the snow beneath. It's not so bad after all (puff-puff), a warm white blanket (puff), a place of security and rest (puff-puff)—and suddenly arrive at the site of another example of this fact. A partridge hole; the place where one of these "wood-hens" denned up for the night, having probably nose dived right into the

snow last evening to escape the cold. This noble game bird, though superbly adapted to snow life, sometimes provides a spectacular example of the term "birdbrain". On occasion the body of one of these avian speedsters will be found lying twisted, but otherwise undamaged, at the base of some great snow-covered rock face, apparently having, in its rush to get below decks for the night, mistaken its form for a depthless drift, with the resulting fatal consequences.

But here, like a little pale crater in a field of white, is the sign of a successful partridge layover. With first light and a growing hunger, the sleeper can almost be seen emerging, perhaps a bright-eyed head first, with a careful scan for danger, then a shower of snow and the bird walks almost regally, if not stiffly away, so completely alert to the many things around it. Or perhaps the bird, sensing the light tread of an approaching fox, or the soft drop of snow as a sagging hemlock branch released its load, came out flying, in an explosion of white and brown feathers. Here signs indicated the former—a soft tunnel perhaps eighteen inches in depth with a fluffy crater lip around its top and a string of rather stubby, three-toed tracks moving off and away. Grouse at this time of year grow snowshoes of their own; tiny fleshy protuberances fan out of each toe giving the bird support in the softest snow and lending a distinctive thickness to their winter tracks in sharp contrast to the thinner "chicken prints" one might find in the spring bogs. No partridge in sight here now, but this certainly does not mean that it is not near, perhaps eyeing me from one of the lumps of snow and evergreen that surround me in the silent woods.

Onward, away from the town. One of the dogs of the village barks and strains against his chain, while up and down the skyline others answer until the bright morning rings with near and distant dog-sound. In trips through the winter woods of the east, a traveler will encounter a distressing number of dog trails, especially near towns and villages where there are, for every confined mutt, three or four that are allowed to roam at will. These often take to the woods.

As a youth, the marshes of northern New Jersey had given me my introduction to feral, or half-wild dogs. Twenty years ago, the vast plains of cattail and phragmites (or foxtails as we'd call them) that grandly flanked the tidal Overpeck Creek, a murky tributary of the larger Hackensack River, were the urban Serengti to wandering packs. Here, within sight and almost sound of the "Isle of the Mannhatoes," wandering bands of dogs made life difficult for the indigenous wildlife and necessitated periodic safaris by the local police when the dogs' boldness, or perhaps an unpleasant incident, indicated a growing danger to children bent on exploring the dense boggy thickets.

It was late spring and I was twelve when I met my first "wild dog." At this time of the year, following the retreat of what snow there is in a New Jersey winter and the rains of April, the "meadows" lay drying under a sun of rapidly growing heat. This was "burn off" time, a time we kids would eagerly look forward to, hoping it came on a weekend. A day would arrive when the firemen of the several towns ringing the marshes would set controlled fires and blacken hundreds of acres of the light brown rustling reeds to remove the danger to human life and property of accidental or purposely set blazes. On this Saturday, the black pall of smoke stood against the sky with flashes of bright orange flame at its base, and summoned every kid within sight, as a monstrous smoke signal, to a day of charcoal and muddy fun!

Trooping down the hill to a scene of a great activity, we would join the annual rite of spring. Like smudged little druids, we bade winter farewell in flame. And after it was all over, with a smokey sun moving down in the west, the meadows lay like a wonderful, steaming prairie. Rough-legged hawks and kestrels hovered here and there seeking out the rodents now deprived of much of their hiding places. It was an almost hellish landscape through which we moved to reach the river: scattered flame, tongues of smoke everywhere and the burnt stubble rather tough on sneakers. As we stood watching the slow moving grey water spotted with the floating, black-curled remains of stems and leaves, we saw the dogs, five of them. They came to drink upstream in an unburned section. And as I and my two companions stared at them in apprehension, we felt some small kinship with these domestic, controlled creatures with a most uncertain but preordained destiny much as our own, that were at this place and time wild, defiant and free. A shepherd, and Irish setter, two mutts of undecided ancestry and a hefty beagle-like dog made up the ragged crew. Legs and bellies black with mud and creek weed, they watched us with bold interest. A dog in the lap or on a leash is so familiar and benign an image that about the last emotion it could evoke is fear. But a dog of glittering eye and intelligence, rough of coat and muddy from travel through stygian woods and bogs is another thing altogether. This dog arouses an ancient dread, sets the body's protective reactions humming, preparing for flight or fight. These dogs, on the bank of the sluggish river, were wolves to us, our unconscious inheritance told us that. We began to move away, whispered encouragements and bolsterings of courage—and they began to follow. Running about, heads down, now up, ears forward, mouths open, tongues out (look at the teeth on that shepherd), could they sense our fear and vulnerability?

They made not a sound and approached us no closer than fifty yards, but I'll not soon forget that serious, intensive trek back; pushing through forests of unburned acres, walking (trying not to run) across still-hot flats of black mud and stubble, and the dogs there—we knew they were there. No barks, whines or whimpers, just this silent, sinister pursuit without attack or harm—a war of nerves. Revenge, perhaps? Or were they looking to us to lead them back to a world to which they knew they rightfully belonged? On and on we went until, at last, the park, the wonderful grassy lawns of the park. The comfortingly familiar flag-pole, the cement grandstands built against the hill, the backstop—the shepherd. He stood alone on the grass and watched me with cool disdain and, no, not arrogance, but perhaps something just below that. I remember the animal's face, long, with intent ears, amber eyes. He moved off and away back into the tall reeds, into a place that today exists only in memory. Buried under tons of fill and garbage and topped with Routes 80 and 95, this boyish wilderness must be relegated to the past, but in doing so, my face stung by cold air and assailed by snow-glare, I see yet another dog. He is no ghost of the past, though. He runs near me here, using the man-made trails of a snowmobile.

He lopes on ahead. He's not my dog, but as he runs ahead, making short joyful forays off into this white world of deep snow and shadowy crevice, he looks back and wags his high plumed tail (he's a husky, a snow-loving dog of the north country) and invites me to share in his adventures. To whom does he belong? What is he doing out here in the woods? Can he be allowed to be so unrestrained and free, will he long content himself with snuffing into snowbound chipmunk tunnels, or will he soon tear the haunches from a pregnant doe should he come upon a deer yard, then to curl up before the fire and dream the hunter's dreams?

But as I follow this well-packed, tank-tread trail of the ubiquitous snow country plaything, the snowmobile, the husky loses interest in this unresponsive human and heads back into town, removing the image but not the thought of just what effect man's intrusion in the winter woods can have. These woods of mine are within easy reach of dogs, and men and their contrivances. These trails penetrate deeply into prime deer habitat, as well as those of other wild creatures susceptible to disturbances such as bobcat, fox and fisher. The snow depth in the forest normally provides an effective barrier to intrusion by four- (and two) legged predators, but well-used snow machine trails, and even those of snowshoe and ski afficianados provide ready ingress to marauding canines and off-season sharpshooters, all to the detriment of wild bird and mammal populations. This is not to suggest that these various sports and activities are bad — not in the least. But an irresponsible pursuit of them *will* have its adverse effects, and in some heavily utilized areas, the effects are all but unavoidable. These are things to ponder when you next find the carcass of a dog-killed deer or an empty deer yard area criss-crossed with the tell-tale man-made tracks.

The deer yard. What is it and how important is it to the winter survival of the resident herd? The term conjures up visions of a corral-like area of trampled, packed snow in a secluded section of the woods, whereas in reality, a typical yard will consist of a maze of well traveled, winding trails that traverse areas of high food-value vegetation. These tracks are well defined, being packed down due to the regular travels of the animals reluctant, under most circumstances, to leave their safety. Once the deer have yarded up in the dead of winter, their

metabolic rate becomes critical. They conduct their daily affairs as close to the pilot light level of energy and activity, as survival in the intense cold places a high demand on energy reserves. Moving about as little as possible, rationing the available forage which may consist of deciduous buds and shrubs and whatever else may be present, the yarding deer are quite vulnerable to any forced activity, such as running through deep snow to escape predators or harassment. The food and browse levels of mid-winter make it nearly impossible for the animals to replenish any heavy expenditure of energy, thus they may later starve from the exertion. No, unrestrained dogs do not belong in the woodlands, winter or summer, for the list of grievances is too long — nests and young of birds scattered and driven out of their territories, deer attacks, harassment of hikers. The list goes on.

And so do I. Shifting my thoughts back to the trail ahead, I note that I've come to the "turnaround", the place where the snowmobiler found the going too difficult for his versatile but not invincible machine. My benefactor in this case, by providing an easy path of travel in its solidly packed wake, the snow machine is an invention of mixed blessing. Indeed, hailed in the early sixties as an end to winter boredom, the snowbelt ORV (off-the-road vehicle) landed on New England en masse. They were everywhere, at all times of the day and night, and bloody skirmishes soon erupted between detractors and defenders, the former claiming disrupted peace and quiet, harassed game, woodland litter, ad infinitum; the defenders citing the joys of the trail, responsible sportsmanship and even assistance in wilderness rescues. But, as with most fads, there came a leveling off point which, to my mind, has resulted in an actual decline in the number of snow machines seen and heard in the woods today. But though, in the face of better regulation and machine design, many of the more obvious objections to snowmobiles have been alleviated, with time and research a few not-so-obvious ones have cropped up. Take the field mouse for example. A broad, rolling pasture is home to many thousands of the little rodents. Under winter's insulating mantle of snow, the mice remain active, tunneling through the soft drifts and dining on buried grasses, buds and their own stored larder. But if that field becomes a playground for two-cylinder revelers, that soft snow, under the repeated onslaught of 400 pounds of machinery and treads, becomes so compacted and dense that mice, moles and shrews are unable to get through it in search of food and soon starve. Because compacted snow remains on the ground longer than the normal snow of the woods or undisturbed fields, those that do survive often cannot dig up and out of the snow cover to reach the growing plants of spring that they so sorely need at winter's end. The resulting, almost complete, elimination of the resident mouse population might sound to some like a blessing but when one considers the many species of wildlife heavily dependent upon those mice for their survival: bobcat, hawks, owls, raccoon, skunks, snakes, etc.; and understands the interdependence of all life in a healthy environment, the scope of the problem becomes obvious.

Plunging once more into untrammeled snow and looking back at the tracks, I can see another effect of indiscriminant snowmobiling damage—to vegetation. On properly laid out and observed trails, such damage is minimized, but where the sport is practiced in areas of wild food plants and cover, shrubs are broken and destroyed, and young trees may be permanently damaged. A man-made Krummholz effect!

Just ahead, a hotel looms against the sky, dwarfing its much less impressive neighbors. Not a place of lodging built of stone and steel, but a home none the less to hundreds, indeed thousands of inhabitants dependent upon its sturdy walls for both food and shelter. A dead pine, four feet in diameter, is a citadel, a marker of importance to this human traveler. Great slabs of bark, curved and drilled throughout with the tunnels and pathways of countless boring insects, project from the snow at its still-sturdy base as though dropped garments. All manner of fanciful forms and action can be read into the scroll-work of weathered bark, the gape of pileated woodpecker workings and the holes of flying squirrel and those left by out-flung branches that remain on the great trunk against the slow force of nature. The crown of the defunct forest giant, actually the top third, lies broken and jumbled on the ground at least sixty feet away. It is said that when a tree falls in the middle of a wilderness there is no sound because no one

is there to hear it, but here one can feel the rush of wind and stinging runnels of rain, hear the shocking screech of rending wood and living tissue and sense the violent concussion that rippled through rock and moss as the dismembered behemoth crashed to earth in that long-past storm. Out of a tragedy such as this, which must happen many thousands of times a day throughout the forests of the world, a rebirth occurs, for in the tree's eventual death, a renewal is given to the woods—space for air and light is opened up and made available to younger trees that had previously struggled in the giant's shadows and among its far-flung branches life finds new shelter and sanctuary moving in to begin the tree's inevitable return to the earth almost as soon as the dust of its fall has settled.

But here, on this cold bright day, perhaps twenty years after the fact, the shell of the big pine stands against the wind, every eroding grain etched in three dimensions, each crevice and cranny finely outlined in snow, a portrait of death and immobility. But beneath the silent facade is all as cold and lifeless as it would certainly appear to be? Yielding to an almost universal urge among naturalists, I rap on the flank of a dead stub to see what happens. Nothing happens—at first. Then, following a second blow with my stout shillelagh (a family heirloom), a whispered scratching at the ragged, broken off tree-top heralds the stealthy appearance of a pair of rounded ears, followed shortly by a grizzled black and white face beguilingly punctuated with a set of bright, black eyes and a wet nose. A sleepy raccoon gazes reproachfully with apparent lack of fear at this rude intrusion. I can only stand foolishly and say "Hi Coon", after which the furry head sinks slowly back out of sight into his woody hibernaculum.

Though this coon was definitely asleep when my unwelcome, reverberating knocking intruded into his warm darkness, he was not in hibernation. Like skunks, weasels, indeed most of the small mammals, raccoons may enter a deep sleep for days, or even weeks at a time if the winter is severe enough, but they do not hibernate. Though the breathing rate slows, body temperature and heart-rate remain the same, or fluctuate very little, depending upon species. In fact, late January and early February is, for racoon and skunk, the mating season. During this cold and dark mid-winter period, it would seem difficult to focus (in the woods, anyway) on much more than keeping warm, but this is when one might most likely expect to encounter these omnivorous nocturnal prowlers abroad, their sexual peregrinations taking them miles in search of the den of a receptive female.

Skunks, in fact, are part of an intriguing interrelationship with that powerful avian predator, the great horned owl. Denning up in the late fall after several days of persistent cold weather, the skunks remain asleep for some four to twelve weeks, depending upon the weather. Then, when instinct calls, the males emerge in February and wander through the snowy woodlands in search of female companionship, a pursuit which makes them vulnerable to the appetite and talons of the owl, which also begins raising a family at about the same time. Since skunks are a staple (smell and all) of the owl's diet at a time when much additional nourishment is needed for their growing broods, the whole arrangement interlocks nicely and serves the best interests of all, even that of the skunks. Females, impregnated in their dens, remain below longer thus avoiding the threat of predation and ensuring a safe gestation period for their developing young.

But back to the tree. It would appear that our black-masked 'coon is the sole vertebrate tenant. No flying squirrel heads appear at the holes punctuating the trunk; not even a downy woodpecker explores the crannies today. But the signs of their activities are there to read, like a forest billboard, by the observant. Here is the deep, oval or oblong hole, chopped out by a pileated woodpecker in search of dormant grubs and larvae. Here the holes and tunnels of summer termites honeycomb the wood, the fine sawdust, resulting from the action of countless tiny jaws, powders the ground (in summer) and fills the tree's empty interior. Sow bugs sleep beneath the bark slabs, and the claw marks of foraging raccoons and bears cover the smooth underbark with natural hieroglyphics. Colonies of ants pursue their complex but rigidly predestined lives within the tree and now sleep in dormancy in its heart. The tree, in death, is both a provider and a storehouse of life, and the woodlot owner with an eye and sense for nature is well advised to allow it to stand until the earth decrees it should fall.

My mid-winter hike has taken me to places that, though within two miles of town, are not often visited by people, their machines or their dogs. Moving down a steep north slope, strewn with great granite boulders and fallen trees that make even snowshoe travel an athletic experience, I find myself approaching one of my favorite woodland habitats. In this secluded little valley now humped with snow-shapes lies the beaver pond. Through the labors of these diligent waddling little engineers, the ecology of this valley has been drastically changed. A leaping, tumbling brook has been made a placid pond and wooded swamp, with all the inherent changes in plant and animal life.

Few other North American mammals have had more written about them — in fact, fancy and in between — than has the beaver. Awkward and clumsy-looking on land, the big rodent is completely at home in the water and cannot exist without it. Its aquatic engineering feats are legend; the dam and canal systems, built to provide living quarters and protection during its logging forays, are often extensive in size and one can imagine the amount of time and plain hard work needed to complete them.

Walking rather gingerly across the thick ice and snow of the beaver pond accompanied only by a light wind and silence, I find it hard to conjure up and express any idea of the energy that has been expended all about me. Once the site is selected, the animals begin dam construction by cutting saplings and arranging them with trunks pointing upstream and anchoring them with stones and mud. Additional material is lugged in until the stream is blocked, and the dam is finished off with mud and stones brought in with the forepaws (not with the tail as tall stories would have it). Following completion of the dam, work commences on the lodge. Branches are piled haphazardly, but so thoroughly that when the beavers are through, a well-nigh impregnable fortress possibly as large as thirty feet in diameter and as high as eight feet is the impressive result. A family of beavers, which consists of three generations, and averages eight to ten individuals under favorable circumstances, assembles a goodly pile of food trees to tide them over the long ice-bound winter. Aspen, poplar, birch and a few other favorites are dragged to the pond and either allowed to sink or jammed into the bottom mud. All of this wood cutting is carried out at a furious pace in fall to beat the arrival of the ice.

Beaver colonies in the vicinity of towns are subject to some persecution, both from humans who may harass, trap, or shoot them, and from roaming dogs that may catch a youngster ashore. The big, naive rodents have no "trap sense" as foxes, raccoons or bobcats do. Their curious habit of utilizing scent posts, much as dogs do, by marking with a waxy secretion from two anal glands called "castorum", often leads to their downfall. Trappers use this scent on trap sets, and the beaver, in its lumbering eagerness to learn the identity of any new intruder, falls prey to the ruse and ends up on a pelt stretcher.

Here in the snow though, there are no traps, stray dogs, or trappers — and no beavers, not up here anyway, for they are all under several feet of the solidly meshed branches of the lodge, or propelling themselves with their big webbed feet through the frigid blue-green world under the ice to and from their larder.

Pushing on past the lodge itself (at its crown a small melted spot in the snow and a soft plume of warm air gives evidence of its occupancy), I find myself recalling the pond in summer: the slap of an alarmed beaver tail, the sun-shot air filled with flying insects pursued by tree swallows and dragonflies, the water's surface astir with the ripplings of trout. Ah, the contrast of the seasons' ponderous swing! It's difficult to convince oneself that everything does not just die in winter, to be reborn again from the frozen leaves and soils, that most mammals and birds don't "go somewhere", to return with the retreat of the cold. No, wait, that last phrase is in error, the correction of which will help explain one of the quirks of the changing season. Cold is not added to bring on the winter solstice; heat can only be taken away, and it is done so, with the shifting of the earth's axis, so gradually that winter's really deep cold arrives long after the shortest day of the year (December 21st), just as the "dog days" of August give real summer weather two months after June 21st, the year's longest day.

So much of our world is water, not only in flowing, liquid form as in this beaver pond, but throughout the soils, the tissues of living plants and animals, and as vapor in the atmosphere. Since water gains and releases heat considerably more slowly than air, the day receiving the least hours of the sun's warmth is but an advance warning, and a slight one, of the coming cold, as the lingering warmth of summer is reluctantly given up by the surface waters, the earth's vital humors.

A ridge lies ahead, not an impressive one, but notable for its rugged character. A granite ledge forms it steep southern flank, its crown covered with a lush thickness of conifers (and one giant tamarack off to the left). On its north-facing incline is a group of fine, old, white birch, so stunning in appearance as they stand bright and lacy against the blue sky. On the piebald trunks of those dead or dying, hardy bracket fungi cling in irregular ranks, their tough woody skins protecting their fleshy interiors against the penetrating cold. Of the many species of bracket, or shelf, fungi to be found in the woodlands, perhaps the best known is a large type known as the "artists' fungus". Favoring the defunct stumps of oak and beech trees, the fungi's underside, or tube surface, is of such smooth, durable texture that beautifully detailed mini-paintings have been produced on it. Properly dried, the hard fungus, with its added decoration, will last many years.

Another durable non-flowering plant is the liverwort. These primitive plants, which look like thin, green bits of leather, grow on tree trunks and rocks and are members of that great group of non-flowering plants called *Bryophytes* (those without vascular tissues or roots). Bright, right, green throughout the growing season, liverworts, tough durable vegetables that they are, can withstand the cold and desiccation of winter well. Their tough "skins" become hard and brittle, olive drab in color, forming a successful barrier against the life-destroying temperatures of January, but swell and grow once again in response to the warming of April.

The ridge has another story to tell. At its western termination, there stands a beech tree of such dimensions that it must have seen many a winter pass beneath its far reaching limbs. Upon the smooth grey bark of this tree, generations of black bears have left signs of their passing tenancy of the land in the forms of long, five-lined scratches. Dogs urinate on fire-hydrants—bears scratch beech trees. But the

Hairy woodpecker on a bear-scarred beech tree

ultimate purpose is the same—to establish a sense of occupancy and ownership, a sense of identity that even we humans are not above, we with our backyard fences and property deeds. The territorial imperative illustrated upon a tree!

Here in these woods in central New Hampshire, bears are present (in enough numbers in Grafton County to produce a kill of over 400 this past season), but seldom encountered by the hiker. It is said of fat men that they are often light on their feet and better dancers than most of the more conventionally proportioned population. So can the same be said of bears. Hefty, big haunched critters, it would seem that a bear would advertise its progress through the trees far and wide, but in fact, unless one can surprise a bear by approaching from downwind, a hiker may pass within fifty feet of bruin and never know it's there. A big (at the time) black bear in New Jersey's Stokes State Forest, gave me a personal introduction to bruins' wariness and reaction to human intrusion. Hiking alone on a late fall day, and making what I thought was plenty of noise walking over crisp-dry leaves in an oak-beech forest, I was suddenly treated to the sight of a startling apparition. Rising from a rather extensive blueberry thicket about a hundred feet ahead of me, was what had to be the biggest black bear in North America. Deeply engrossed in adding the withered remains of the summer's blueberry crop to its winter fat layers, the big fellow had been caught unawares, stood up with a loud woof, and was now snuffing and squinting about trying to get a "fix" on me. In another second he did, and came down running—in the opposite direction from that of my own flight, which was well under way. Seeing that pursuit was not this bear's intention, I stopped and listened to his swishing, crackling progress as it faded slowly off through the trees, leaving me with a pounding heart and a memory that would stay forever with me—my first and only meeting with a truly wild bear in his own world.

This bear-scratched beech tree, with scars of varying freshness, indicates that an occasional animal at least passes through the area during other times of the year. The spot is too close to town to be a denning area, though with its maze of rock crevices and exposed tree roots, it would make a nearly ideal one. Within such small caves, *Ur-*

sus americanus might have, in centuries past, slept (not hibernated) the winter away. But perhaps in place of them we might scour up a porcupine. These prickley, orange-toothed rodents, of lumbering gait and seeming stodgy and humorless manner, are woodland dwellers of wide-spread distribution, though not often encountered by the casual traveler. Its scientific name *Erethizon dorsatum*, is derived from the Greek verb, meaning to excite or irritate, a reference to the formidable spines. Indeed one can think of nothing else when the porky is mentioned, its quills are as much a part of its image as a skunk and its smell.

Porcupines have a positive need for trees. When not denned up during extreme cold weather, they are aloft in the nearest hemlock or black cherry or intently plodding between the two. They have a never-ending craving for salt — according to J. Kenneth Doutt in "The Mammals of Pennsylvania": "Nothing is sacred — camping equipment of all kinds which bears the slightest trace of sweat is eagerly gnawed to destruction, even the floors and doors of the cabin itself. Cupboards are gnawed through and the contents ruined as this rodent systematically seeks its favorite flavor." In most cases, the "quill-pig" is content with the occasional discarded deer antler found on the forest floor for its seasoning needs, but in areas of over-abundance can be (according to our White Mountain foresters) damaging to commercially valuable trees by girdling them or stripping them of too much bark.

Here in a sheltered rock overhang and tunnel, found by following converging pigeon-toed footprints from a porcupine's frequent trips to and from its food trees, I'm lucky enough to find an occupied den site. Fresh droppings strongly resembling those of rabbits, solidly pack the entrance way, a greenish-brown carpet of odorous buckshot, receding into the blackness. Leaning close, I wonder whether the faint sounds I catch are the drip of subterranean water, the furtive scuttle of mice or the uneasy stirrings of porky himself. Perhaps he is backed up against the inner wall of his hideaway, seeking, with his myopic, rather witless eyes, the nature of the puffing, noisy creature at his door. Porcupines can grunt, chirp, and squeal when the occasion demands and, when disturbed, can also deliver a strong-jawed bite. So, rather than press the issue, let's look further about the ridge for signs of winter life. Tracks abound in the snow; the meandering big-footed trails of snowshoe hares transect the hemlock and balsam groves. Birds are active today, too: a downy woodpecker; there a brown creeper working its way up an elm trunk. These sepia-colored, solitary little birds of northern woods move from tree to tree, always landing at their bases and moving upward to the higher branches, thus moving about, one might say, in great letter N's. The presence of winter birds in any numbers indicates a wild food crop able to sustain them through this difficult time. Though the term "wild food" usually refers to seed-bearing plants and the fruit of shrubs and trees, many species, such as the popular chickadee, depend upon sources of dormant insects, their eggs and larvae. These two are wild foods, and the numerous, perhaps unsightly but so essential dead and dying trees that stand scattered all across the ridge among their still-living brethren, are veritable supermarkets of invertebrate abundance, the natural feeding stations and the staff of life for the birds.

With the sun well on its way to the horizon, shadows lengthen, imperceptibly swinging across an almost rainbow-hued snow cover. A maze of pale mauve-blue lines intersect and flow over the endless contours of the forest floor at this, most peaceful and evocative part of the day. New tracks are etched against the snow, sought out by sun and shadow. A bobcat has passed here, though from the indistinct nature of the prints, single line and regularly spaced like all cats, he came though during the previous night. The trail of *Lynx rufus* looks like the track of an overgrown tabby, albeit one you wouldn't wish to stroke on impulse. For though the wildcat is one of the most beautiful of North American mammals, it is also admirably equipped for close range combat and is not in the least a social animal. This particular cat, the author of these tracks, doubtless has a denning site nearby and is reaping a harvest of snowshoe hares, just as its kin further south of the winter woods of Pennsylvania or New York might exploit the cottontail, carnivores that have consistently suffered from a bad press and public image. Though of fearsome temperament and appearance (who has not shuddered at the sight of a wildcat of any species snarling and spitting its frustration at the world from behind the bars of a zoo cage?), the bobcat is a relatively unaggressive animal that will not attack unless cornered or wounded. Wolves, skunks, spiders, snakes, hawks, and owls, all are creatures misunderstood, feared and unjustly persecuted, and primarily for dong just what we do for sport or need — hunting.

Evening, and the renewed cold of night, is stealing through the forest; it has grown so queit, no wind moves the branches or sings through them in wintersong now. A flash of white on white ahead betrays the snake-like motion of a hunting weasel, he in his white "ermine" (the robe-trimming of royalty) pelage as he flashes here, now there; up and over snow covered rocks and roots in a never ceasing quest for warm living prey. Today I've been fortunate for one can spend hours, indeed days, hiking the winter woods seeing nothing but indirect signs of the wild creatures' presence. The animals usually manage to be aware of all but the softest approach and absent themselves quickly, long before our eagerly searching, but civilization-dulled, senses arrive on the scene!

The light is failing rapidly now. Snow crystals on the tops of snow banks and the stalactites of ice festooning trees and rock faces catch and flash the mellow rays of the dying sun. Blazing beauty born of cold. At the manicured edge of the park some of the few goldenrod stalks of summer that thrust above the snow sport galls, those globular insect hideaways produced by the saw fly. These flies are equipped with a "stinging" apparatus by which they inject a substance into the tissues of the particular type of plant parasitized by their species. This substance causes the immediate area of the plant to swell, as we might do upon the receipt of a bee sting, and into this enlarged area, the parent fly lays an egg which hatches into a larvae. This lives within its snug home until development decrees that it must gnaw its way out into the world.

Due to the fact that there are thousands of different galls produced by insects, some on stems, others on leaves, twigs, flowers, etc., it is an exercise in futility to attempt to give the particulars of appearance and life history of the phenomenon of galls. The goldenrod galls are perhaps the most obvious in the sparse winter landscape and, upon closer scrutiny, some will display tiny, neatly drilled holes, indicating the tenant has matured and has left to seek its fortune. Those unblemished individuals will hold, as a cross-section cut with a sharp knife will show, a tiny white worm, curled snugly within a cellulose cradle. Another commonly seen gall, readily noticeable among the bare branches of winter trees, is the blackish, lumpy "growths" seen on the twigs of fruit trees. The products of the labors of the saw fly, callirhytis, these excrescences provide shelter for the pupating grubs of this fly, and look something like bits of burnt marshmallow stuck on their respective branches. They appear to be most often seen on wild cherry twigs.

After spending the day on snowshoes, wading through light fluffy powder that grows imperceptibly heavier by the hour, one's feet feel light as goose down at day's end. You can almost feel thigh and calf muscles singing, aching and growing in strength and endurance. Crossing the brook again, I long to drop down and take a long, luxurious drink out of the few ice-rimmed areas of the channel still open, but the words of an old woodswise friend admonish me: "Never drink real cold water or eat snow after you've been walking all day in the woods and are sweatin' hot. Give you heart failure."

True or false, I don't know but I opt for passing it by as I'm within a half mile of home and nice glass of drawn water, or better yet, a beer. Pushing on, I argue with myself. Am I a traitor to the woodsman's way, afraid to take a drink? Or am I being justifiably wary; is there fecal waste or God-knows-what in this clear brook, given its proximity to civilization? A lethal dose of coliform bacteria in every draught?

Another owl interrupts my reverie and self flagellation — shadow form, and silent flight. A small owl leaves its perch in a bushy stunted balsam and glides a short distance to an exposed branch on a bare maple. Small, somewhat larger than the elfin saw-whet owl, the bird sits stolidly on its perch regarding me with yellow, unblinking and unsuspicious eyes. A quick scrutiny of its most apparent field marks (no ear tufts, streaked belly and short tail) identify it as a boreal owl, a

Goldenrod galls

rare find, indeed. This is a secretive gnome, the "water dripping owl", so named by the early Indians because of its liquid repetitive call notes which recalls the sound of falling drops of water much like the irritating midnight noise produced by a persistently leaky tap. The boreal owl is a true bird of the northern forests. The appearance of these, or any of the other arctic or tundra owls like the snowy, great gray, or hawk owl, surely prophesises a hard winter.

The owl's head bobs and turns on its short but supple neck (fourteen neck vertebrae as opposed to our seven), then snaps to attention as it gazes intently at a slight distant movement that I cannot pick out. Leaning slightly toward the object of interest, as if to confirm its value as food, the bird hesitates for a mere second and then soundlessly leaves the branch, defecating as it goes, into the gathering dusk.

Once again, the branches of trees, set in motion by a freshening breeze, cut and bisect the blinking lights of town not far ahead. Reaching the street, snowshoes, which seem to weigh ten pounds each by now, are removed. The road stretches ahead into shadow; packed salt-filled snow and sand spread by the town road crew crunches like Grapenuts underfoot. With spring runoff, all of this crud will find its way, alas, into the little brook nearby, to the discomfort, I'm sure, of all the gilled creatures that live there in it.

A quiet group of drab house sparrows perch sleepily in the tree near our feeder, feathers fluffed against the cold. Hardy little birds, urban freeloaders (I once watched one hop through the grille bars of a '72 Chevelle and pick the bugs off the radiator!), house sparrows are with us on the worst days of mid-winter, when nearly all others are out of sight. They cause me to remember a poem by Leonard Cohen:

"Catching winter in their carved nostrils
the traitor birds have deserted us,
leaving only the dullest brown sparrows
for spring negotiations."

I find I too must turn traitor and seek the warmth of the house again, though I, like those beings I've left to the now dark and still woods, begin to look again toward the days of April....

LATE WINTER
A Time of Hope

*In March the door of the seasons first stands ajar a
little; in April it is opened much wider; in May the
windows go up also; and in June the walls are fairly
taken down and the genial currents have free play
everywhere.*

—*JOHN BURROUGHS*

It rained early this morning—a cold, later freezing, driving rain,
but the first real rain of the new year. Along the sloping lengths of
telephone lines, the starlings of the town and the few early redwings
and grackles that joined them, sit in wet ranks, soundless and forlorn.
Few chuckles, whistles and any of their other varied calls do they give
on such a morning. Ragged scraps of pale blue begin to appear low in
the west, and by mid-morning, as I prepare to take an early spring
look at "my woods," a dry, warming wind moves under a lifting overcast.

The woodlands today appear as a widespread of leafbrowns, rich
greens of recovering mosses and ferns, and the ragged white of the
retreating but still very much present snow. The leaves and last year's
grass in the yard have a curious flattened look about them, as though
the winter snow cover had bore down upon the ground with a great
weight. On this compressed, rain-wet compost a solitary junco hops
fitfully about, flicking its white outer tail feathers in what must be a
fruitless search for insect nourishment. A lone jogger appears on the
road, running resolutely, face red and set in grim healthful deter-
mination, grey sweat suit plastered to his chest and legs, hood pulled
tight transforming his features into a bright oval.

Surveying this cold and still forbidding early spring scene, it's

Chickadees at a suet feeder

hard to imagine that further south gardens are being prepared, spring peepers are filling the night air with sleigh-bell sounds, and infant leaves are emerging into the warming air.

In the woods the snow is leaving! The strangely beautiful substance, so stainless white and of such evanescent texture that graced the forest floor of January, transforming each evergreen thicket into an arctic jungle inhabited by bright white giants, has receded into debris-spotted, ever shrinking patches. These too will soon yield to the coming rains. The time for snowshoe travel is past, but every rushing brooklet, every green patch of ground pine, each bursting bud proclaims that the time of new growth and awakening is upon us. The season of snowbound nights and cutting winds is done—spring is in every scent and sight today.

Among the trees, it is possible for one to forget oneself, time and space. In every direction, the snow-shapes of winter have yielded to a heady, scented dampness, an indefinable odor consisting of eons of leaf mold, the pungent scent of new skunk cabbage and cinnamon fern—a physical stimulation that supersedes any man-made intoxicant.

In the park, filled with mini-lakes of snow melt, the resident red squirrels are noisy and active. Here they are the dominant species, forcing the larger but less aggressive greys into the town with its cropped lawns and maple lined streets. Red squirrels, spunky little rodents, don't adapt well to a people-oriented environment (you'll rarely find tamiascurus sporting about in an urban park). Lovers of the evergreen treetops, these brick-red little dynamos are most often encountered in areas retaining at least a trace of wilderness. Five hundred years ago, these "little tree rats" probably regarded their human neighbors in much the same way that they do today—as unmitigated nuisances and intruders to be on the receiving end of considerable squirrel cussin' and swearin'.

It is said of the red squirrel that it will tolerate no brooking of its territorial boundaries, and its loud shrill chatter will greet any intruder, large and small. Legend has it that the excitable little creatures have a deep enmity for their larger cousins, the gray squirrels, and will chase, catch and castrate them (the males at least) at every opportunity. Though many people with even a modest degree of woods experience have seen an infuriated little "chickaree" pursuing a big gray through the treetops, no recorded instance of such a bizarre attack seems to exist. Actually the two species live in relative harmony, with the gray squirrel appropriating the better habitats, often relegating the red to the fringe areas. One would suppose that this 'second class citizen" status would instill a resentment in the smaller animal and account for its noisy, aggressive behavior, but I think our little "pine squirrel" just has a strong territorial sense and would defend a zoo cage with the same ferocity!

On my way through the park this morning, I saw another squirrel, this one a true creature of the forest, the enigmatic flying squirrel. Though nocturnal in habit, these soft little sprites will make an occasional appearance on overcast days, as did this one, scurring up a dead stub and vanishing into its den hole. A brief glimpse, but one, like all little happenings in nature, to be treasured and remembered. Our particular flying squirrel here in New Hampshire is the northern race, *Glaucomys sabrinus*, more robust of build and brighter in color than its southern relative. Some confusion is likely where their territories overlap, thus the fine points of differentiation are best left to the taxonomist examining skull structure. Though widespread in distribution, and reasonably common in most forests, the presence of flying squirrels very often goes unnoticed due to their preference for the hours of darkness and their secretive nature.

The breeding season having been completed by late March, this individual may well be an expecting female preparing a home base for her impending brood. Upon arrival, the young are pink, hairless and blind, but show the characteristic gliding membrane, the fold of skin between the fore and hind legs that sets this little squirrel apart from all others. The flying squirrel does not, of course, fly in the true sense of the word. It glides or volplanes from tree to tree, taking care to avoid landing on the open ground or other such exposed area where it is less agile and thus more vulnerable than the larger squirrels that spend time on the ground.

At the Squam Lakes Science Center in central New Hampshire, where I once served as artist-naturalist, we maintained a colony of these crepuscular gremlins in an exhibit called simply, "The Nocturnal World." Here, in a roomy "territory" of dead branches, carefully carpentered nest boxes and abundant food supply, the luminious-eyed rodents conducted their lives with seemingly innocent abandon, capering about in a "night" of infra-red light during our day, and sleeping under the glare of light bulbs at night. This rearrangement of so strong a behavioral trait was not the easiest thing to accomplish, but with time and patience, it can be done. In this case, of course, it was done to facilitate viewing by visitors to the center's exhibits.

The snowmobile trails of January now lie flung through the woods in great white meandering ribbons against the wet leaves of spring. Their packed, well-traveled paths will be the last snow to disappear from sight as the rains and returning warmth signal the winter's end. For almost a week in advance of this gray day, it seemed that rumors were afloat in the air. One could sense the turning tide of seasonal change. It could be seen in the sun, higher now in the southern sky, and in the length of the day. Suddenly, shaggy-barked maple patriarchs on country roads, in backyards and on farms have sprouted sap buckets like grey metal fungi, the larger trees yielding the quickening sap flow to as many as four. Sugarin' time in New Hampshire is as sure an early sign of coming spring as the ice-out on the lakes. There are many sugar maples here in the woods as I pass through on this walk, but most of them are rather spindly and crowded, not "sugar bush" material. Sugar bush is the New England term denoting any stand, no matter how spread-out, of maples from which the sap is taken, and the harvest is, in the words of John Burroughs, "the sweet goodbye of winter."

The sap begins to run just about at the vernal equinox toward the end of March, and conditions for a good run are exact. Daytime temperatures should not rise higher than forty degrees, nor those at night time fall below twenty-four degrees. Old timers say that a northwest wind is best; a warm south wind will end the run, temporarily at least. Maple sugaring is a purely American pursuit, dating back to

Flying squirrel

early New England, and to the Indians before. The colonist collected the sap in rude wooden troughs and boiled it down in kettles suspended over hot fires. Later, saphouses were built for the purpose. These odd-looking structures with their distinctive ventilating cupolas on the roof, stood silent in the woods for nine-tenths of the year, then burst into smokey, bustling activity as the sap season arrived.

Moving through this especially open section of forest, where even the pallid light from a cloud-hidden sun provides some radiant warmth, I encounter a strange sight, The snow here, rumpled and dirty with spring-melt, seems to be undergoing strange optical contortions. A sort of grainy film is collecting in every available low spot and shadowed recess so that depressions and ridges appear to be rippling and in motion. On closer inspection, this illusion resolves itself into masses of moving black specks that jump and vanish before the eye. They are snow fleas, those strange looking little oddballs that appear on early spring days of bright sun and cavort in apparent celebration on snow, water surfaces and even, to the dismay of the sugar bushers, in buckets of maple sap!

The snow flea, or glacier flea as it is called in the west, is one of various wingless insects called springtails. As scrutiny under a microscope will show, it is a grotesque little creature that gets about in the world by means of a unique mechanism quite different than that of an animal flea, to which it is totally unrelated. Big bodied creatures, with ineffective legs and a large head, springtails use their highly specialized caudal appendages as a sort of coiled-spring device. Normally it is kept tucked underneath the body, but when the flea wishes to vacate the premises in a hurry, the tail is rapidly straightened out. Presto! The bug is air-borne! Multiply this simple act by a million-fold on the snow of a bright late-winter day, and you have the snow flea phenomenon.

Springtails are among the most primitive of living insects; at least as old, geologically, as the obnoxious silverfish of book-and wall-paper eating fame and the wily cockroach. Other insects, most of more recent origin, also begin to make themselves known outdoors in late winter. On warming days, whirlagig beetles will appear in numbers, tracing their idle circles on quieter stretches of open water, often right behind the retreating ice. Like miniature turtles, these steel-blue, metallic looking insects can also be seen sunning themselves on plant stems and, in more southerly climes, will come out of hibernation in January thaws for mid-winter dances. Though dependent upon the water world for food, protection and reproduction, whirlagigs can fly quite well, but cannot launch themselves into flight from the water's surface since they need an elevated takeoff point such as a twig or reed stem for the attempt.

Below the surface of the freshening brooks, mayfly larvae, as well as those of the stone fly, begin to stir under rock hideaways. Soon they will leave their aquatic existence behind forever, crawling out onto the stones and logs of the home brook or pond. Here the fully grown nymphal skins split and the adults emerge. Like a departing invasion force, they take to the air, leaving their former skins firmly hooked to the launching pad by their now empty legs!

The annual insect arrival, commencing when winter's white still covers all and reaching a peak with the not-so-welcome appearance of the mosquito, "no-see-um", and black fly hordes, occurs on such a small scale, that few people realize that, in scope and activity, it is probably nature's greatest show on earth. Outnumbering all of the world's other groups of inhabitants (both in number of species and individuals), the insects are everywhere. A mere handful of earth and leaf litter may contain eggs and larvae in the thousands and every nook and crevice of a brook, any given square inch of pond mud, harbors a miniscule society engaged in as serious a struggle for survival as any conducted at mankind's level.

Look down today; scrutinize the singing torrent in the brook, rushing between walls of snow and ice-coated rocks. Here a few sluggish back swimmers jerk about in search of tiny prey. There, in a quiet eddy, a caddis worm hauls its tiny house of sticks across a smooth mud-bottom yet unblemished by the tracks of other, more warmth-loving creatures. Where the water fans out into the low spot (where partridges stalked in early winter) torpid dragon fly larvae cling to the dead stalks of yesteryear and await less hostile days and temperatures

Boreal redback vole

in which to complete their life cycles. In little more than a month the air here will be busier than any airport—hover flies, wood satyrs, darning needles, crane flies—the list is almost endless. But for now, Nature, a great and ponderous mother, will move slowly, presenting us with such tidbits of active life as can be seen here today and saving her grander presentation for May.

Even now, in some snug retreat below leaf and rock, or perhaps deep within a cave in our dead tree, a fertilized queen bumblebee begins to stir in answer to the imperceptibly lengthening daylight hours. Even before Jack Frost's touch has been lifted for the year, she will venture forth, like a tiny bird, in search of a suitable site in which to prepare a home for her soon to arrive subjects. This may be an abandoned field-mouse burrow, a hollow log, or even a bird's nest of last summer. After arranging dried grass and small twigs to form the actual nest, the bee collects pollen and places it in a pad on the floor of the nest. Upon this pollen-mass she will lay her eggs, covering them with beeswax and brooding them like a hen during cold weather. After hatching and undergoing a ten day pupation period, the first bees emerge, workers all, and take up the house-keeping duties, thus introducing yet another generation of *Bombus* to the world.

What other signs might I find today? Insects, spiders beginning to move, the hidden potential of *Cecropia* and mantis cocoon perched on twig and briar. At the pond, snow and ice on the bank chips and crumbles. One can almost hear the life stirring beneath the mud and bottom ooze, though as yet few fishes are to be seen. The pond's water has that fresh new clarity that it seems to possess only in early spring. Later, algae bloom, lowering water levels, and pollen debris will all contribute to a general murkiness that will reach its apex in August. For now though, it has a crystalline, cut-glass quality, and suspended in it moving about in jerky animation are newts, vanguard of the am-phibian choruses that will soon fill the chilly dusk with ringing, clack-ing love songs. Today, though the air has that warming, expectant quality about it, one can still feel the refrigerating presence of the snow. In the painfully (to me) cold water, the newts are carrying out their amorous courtships which makes for fascinating observation.

The male, identifiable by his much larger, ridged tail, has seized the object of his affection just behind her forelegs. Holding tightly, he now bends his body into an S-curve and begins rubbing the side of his head against hers, all the while gently tapping his tail against her body. Twisting and turning, completely oblivious to prying eyes, the animals may perform thus for hours until, upon separating, the male deposits his tiny vase-shaped spermatophore on a stone or submerged leaf. In a "newt-pond", hundreds of these white objects will be seen scattered over the bottom. The female then crawls over the spermatophore, takes it mass of sperm cells into her cloaca, and then deposits her fertilized eggs, one at a time, in her subsequent travels.

The results of all this is a tiny, gilled larvae, which, at the end of the summer, has developed into either the familiar green and red-spotted newt that most children have tried to keep in a jar or coffee can, or the translucently beautiful red eft or land form, which is best seen in the woods after a rain. What percentage of a pond's young newt generation will remain in the water and what will take to the forest nearby is one of those little mysteries of nature not fully understood.

The presence of the amphibians, frogs, toads and salamanders, is one of the more obvious, exciting events of spring. Long before the first wood thrush utters a chirp from the glens, before the woodchuck sentinels appear at their roadside salad bars, the ponds and bogs ring with frog sound. It is as though a great convocation of all the gnomes and trolls of the forest has been called; from their winter sleep in subterranean dens, they appear in their ancestral ponds and give vent to a joyous welcome to awakening life. The sleigh bells of *Hyla crucifer*, the spring peeper, the clacking din of wood frogs, the trills of toads and the twanging plunks of the green frog all merge into one living sound that few country folk have not marveled at!

Prowl the spring bogs at night with a flashlight and you'll get a glimpse of the singing multitudes before they once again vanish into the meadows, woodlands and swales that are home for them at other times of the year.

Not far from here lies a shallow body of water, not fifty feet across, that all but evaporates into the air in August. Huddled in a remnant of pasture, in a triangle formed by a shopping center parking lot and two well-traveled roads, the pondlet provides a perfect example of what can be found in even the most restricted and tentative environments by the sensitive observer. Snipe cling to the flooded alders at its margin and killdeer run over the rolling pasture. Frogs and spotted salamanders crowd the water, braving the "no-man's land" of asphalt on all sides to get there.

Exploring the spot one early April day last year, I discovered the egg mass of a spotted salamander (the eggs are larger and farther apart than frogs eggs) and resolved, as not one was in sight, to return that night. The spotted salamander is one of our largest and most attractive species, being a deep lustrous black with two rows of round yellow spots from head to tail, and reaching the impressive (for a salamander) length of six to eight inches. Nightfall found me back on the scene clad in waders and heavy jacket, for the night air was a chilly thirty-eight degrees, and wielding the necessary flashlight. Probing the black water, the beam soon picked out my quarry, a milling ball of at least two dozen "spotties" engaged in procreation. An animal this strikingly patterned and colored, against a light background of submerged grasses, would stand little chance of escaping predation if it pursued its lovemaking during daylight hours. Great blue herons, bitterns and other diurnal predators would make short work of the revelers under those circumstances, not to mention school kids and terraria collectors with nets. Nature has decreed that the salamanders, indeed most amphibians, pursue their reproductive activities at night when they are far less easily discovered and devoured.

Though New Hampshire's amphibia will not, under usual conditions, make an appearance until mid-April or even later if frost is persistent, the first week of March will often ring to the spring peeper chorus in the winter woods of more southern regions. Snow cover is very often entirely absent there and ice does not attain the thickness it does here (often twenty-four inches), so that a warming day will bring these little frogs out of hibernation at an early date. On one occasion, a friend and I collected, with the aid of a flashlight, some fifteen of the little sprites in order to observe them in a twenty gallon terrarium I had set up at home. Introduction of the peepers to their new, attractively landscaped home was uneventful and all went well until about midnight. We had just retired when a solitary singer began his song, and my wife and I lay in bed and listened with pleasure to this concert of nature right in our own home. Our joy was short lived. After perhaps fifteen minutes, our soloist was joined by another, then another, another, another—until our modest apartment positively vibrated with frog music. The kids began to stir and neighbor voices upstairs were heard. My wife sat up and just looked at me until I crawled out of bed and covered the tank with two towels and a quilt. The little brutes quieted down immediately and peace reigned for about five minutes when the racket began anew. We felt as though our bed were afloat in the middle of a pond! Dogs began to bark outside and the whole town seemed in danger of being awakened.

Finally my wife's malevolent glare compelled me to act. Resisting a terrible urge to hurl the frogs, tank and all, through the living room window, I lugged the heavy aquarium, aslosh with water, ferns and the frogs down to the basement where they peeped away till dawn with only the spiders and washing machine to hear them. At eight o'clock in the morning I gratefully gave them their freedom, resolving to do my peeper field work in the field from then on. Perhaps this is what they had in mind all along!

Ah, the voice of crows! In late winter they are all voice, seeking each other out, arguing in the tree tops, having fun before the serious business of raising a family begins. Walking up the logging road toward the rock ledge, I am assailed by crow voices floating over the treetops and echoing through the valley. Winter is ending, they seem to shout, hard times are done! "Hardy, happy outlaw, the crow, how I love him!" exulted John Burroughs of the adaptable bird. When it comes to surviving all that this difficult world has to throw at you, certainly the crow is a good survivor—he has weathered it all. His voice has not the sweetness that Aesop's fox attributed to him, but it serves him and his tribe well. Far carrying and expressive, it is a garrulous link between all crows in the vicinity. An owl is found. Danger is near.

Crows

Food! All is well. Such messages are conveyed between the black brethren as distinctly as though by telephone. How much character there is in a crow's call!

Approaching the ridge, I can see the author of but one of the raucous cries. He sees me too for a crow misses nothing and allows no furtive approach. Somehow he knows I have no gun and so stretches and postures like a rooster with beak wide and throat fluffed and sends forth his lusty call across the red and grey mountainside. But as I approach closer yet, wisdom prevails over arrogance, and, dropping from his high perch he sails away through the trees, braying defiance as he goes. The world will never see the end of crows, unless it becomes unfit for all life.

On the logging road, fresh-turned earth, lumps and wandering lines of it, indicates moles have been at work through the winter. Insectivores like the superactive shrews, moles do not hibernate, but pursue their annulated wormy prey deep into the innards of the earth, retreating with them from the touch of the frost. Anyone who has ever worn a moleskin glove has felt the mole's secret for successful underground movement. In its pursuit of either food or escape, the mole must often back up as well as dig forward. To facilitate this, its fur has no definite "lay" to it: it may be brushed forward as well as backward. This allows the mole to change gears in "mid-dig" and back up without resistance to avoid a burrowing dachshund or the spade of an irate gardener.

Moles are not as common in the deep woods as they are in pasture land and lawns but one can often discover the meandering tunnels of one of the tireless escavators crossing a sylvan pathway. Mole tunnels are evidenced by those lengthy ridges in sod that traverse lawn and field alike, while the proverbial molehill, the result of the excavations of a female building a nest chamber, appear on level, soft ground in the spring. Three species of moles, the common, hairytailed and star-nosed, are denizens of the winter woods, and signs of their presence and activities can be expected and seen as soon as the snow leaves the ground.

The rock ridge, denuded of snow now, reveals something that I was until now unaware of: an old dump site and the timeless erosion of the rock itself, two indicators of time's slow march. The ridge is less than 100 yards from the cellar-hole ruin, so I suppose it should not have come as a surprise to find the dump out here among the trees. Though I had many times explored the remains of the house, I somehow overlooked the fact that the inhabitants, though discarding little in that thrifty age, must have used some spot for throwing out whatever trash they did accumulate. Though still partially under snow cover, a treasure trove of rust-eaten stove pipe, old sapbuckets and metal sheathings along with a wide assortment of unidentifiable farm implements is revealed. Disintegrating metal cans, their tops opened in the ragged circle made by some primitive can opener, suggest that the homestead was likely occupied up to the turn of the century, perhaps abandoned by a new generation in favor of something bigger, better and more civilized. Since the old foundation is clear, save for natural debris, the other elements of the building must have been scavenged part and parcel and perhaps put to use in the building of its successor, leaving only an empty cellar to the mole crickets, wood frogs and garter snakes. Such a place provides a somewhat unsightly, but secure haven for the reptiles and many other woods creatures. I well remember finding a king snake, resplendent in bands of russet and pale brown, under a sheet of corrugated tin in such a dumping ground several years ago. Taken home and placed in what I thought were adequate quarters, the snake managed to make good its escape several days later much to the acute dismay of my mother-in-law who was visiting at the time and who has a definite lack of appreciation for serpents. A subsequent thorough search of the premises failed to turn up a trace of the escapee, so it was hoped (fervently by her) that my pet had found its way back to the rocky habitat from whence it came.

Here in the woods, covered with the glitter of the spring melt, the stoney faces of the ridge stand subject to the slow erosion of another year. Weathered rocks have such a tale to tell. Here and there through the woods they lie—great rounded boulders, travelers perhaps from far places, dropped and abandoned by retreating glacial ice to witness the passing of mastodons and early men. Or perhaps it is the protruding backbone of the land itself, this very ledge, once an ancient landscape, now creased, flecked with glittering fragments and lifeless. How long it takes for mountains to fall and the very foundations of the earth to shift and move. The comprehension of geologic time is all but beyond our grasp as we, like most organic life in a given present, pursue our short lives. But the rocks know it, understand it, for they will exist, though perhaps diminished in size, when it ends. Place your hand on the flank of the ledge and try to take from it some sense of the span of existence. Time is the enemy of rocks, for it brings with it powerful allies to aid in their destruction. The rain washes away the protecting mantle of topsoil and with patient touch, the process begins. Lichens and a half a hundred species of liverworts anchor upon its surface; debris gathers and the seeds and spores of forest plants jostle for a foothold.

Autumn's flow of water becomes winter's rigid wedge, exerting a slow but irresistible pressure that widens a fissure enough to permit a tree to root there; its growing and spreading root system gathering strength from the soil, the sun, and time to further separate and crumble the rock.

In the brook one can see a graphic illustration of the ultimate fate of the rocks. The current drops it heavier contents first, so that stones make the bed of the rapid flowing brook, while near the outlet to the pond, where it courses with decreasing speed, pebbles and then sand bottoms appear. Finally, at its delta, little but fine silt is carried in suspension to the pond's waters, to be spread far and wide over its bottom, building a nursery for new life. But still the weathered rocks endure. They are friends of man, in that they tell us of our planet's past, form our buildings, and the monuments that stand above us when we pass on. They also resist us in their passive way, making the land untillable by their presence, so that they must be moved. The old farmers built fences of stone less to keep their animals in than to get the rocks out of the way.

Rocks, "the bones of the continents", have had a long and intimate partnership with man. Primitive men expertly chipped arrow heads and spear points of quartz, while flint, struck and sparked, has kindled all manner of conflagrations, from rude campfires to Zippo lighters. Rocks and their shadowed recesses have provided our forebearers with places of safety and shelter, and our children with imaginary ramparts of play and adventure. Annointed today, they sparkle with glint for all the world as though in celebration.

The beaver pond too has a radically different aspect than in mid-winter. Though still showing no active signs of the owners' sence, the ice is beginning to draw away from the shore in places, and the lodge itself has the appearance of some great white pin cushion, with its myriad logs and sticks poking up through the shrinking snow. Though I can neither see nor hear them, I know the beavers are there; perhaps last year's kits, approaching a year in age and with it, a growing independence and desire to set out on their own in the world, are huddled with their parents within the musky dark interior of the lodge. As soon as the ice disappears, the amphibious creatures will venture forth again, and in May, a new generation of *Castor* will make their debut, chirping and wobbling about in baby-clumsiness snug within their log and mud fortress.

I dare not try the pond's ice covered surface now. Slushy and leprous-looking, with a somewht greenish-yellow cast, it will not, I am certain, take my weight, and thinking back to a "breakthrough" of my youth, I decide not to attempt to reach the lodge. That past incident had occurred at the edge of the vast tidal marshes of the Hackensack River in New Jersey in late February when I was an exploration-prone fourteen-year-old. Prowling about in search of wintering ducks, I was startled by the sound of a malevolent voice not far away: "Get outa here you little S.O.B., you're on private property!" it roared, and looking up, I took in the imposing figure of a rather elderly gentleman clad in some sort of uniform and rather aggressively cradling what looked to me like a Gatling gun. Patrolling the edge of what seemed to be a landfill site, the man was obviously in no mood to discuss the pros and cons of the situation, so effecting a timid wave, I beat a rapid retreat in the opposite direction. Crossing a frozen tributary of the river, I neglected, in my haste, to note that the tide was out, and though the ice remained stationary, supported by its own rigidity,

there was about two feet of empty air between it and some extremely unattractive mud below. Upon reaching the near mid-stream point, the ice gave way and with a whoop of dismay, I dropped to the quagmire below with the ice itself nudging my armpits.

Trapped in this unfrequented place, I could easily have been forced to spend the night in an awkward position. My violation of the first rule of winter wilderness travel — don't go out alone — had put me a dangerous five feet away from dry land, which looked like five miles! As it turned out, some energetic wriggling, huffing, and puffing did the trick and I managed to free myself and crawl to safety. Reaching home an hour later, my muddy and soaked pants frozen stiff against my numb legs, I had ample time to reflect upon the treacherous and misleading qualities of winter ice.

Best to skirt the pond today and investigate the natural signposts of spring's approach in pathways of safety! The slow but noticeable motion of the water under the ice in the pond seems to create conditions favorable to the early revival of life. Watercress, a common and edible aquatic plant of the woodlands which remains a bright green through the coldest winter, is beginning to advance, spreading and branching out against the bottom mud and leaf litter seen through the water of the pond's outlet by the dam. Here, the water's flow has kept the ice at bay through the winter, and with the warming, it retreats quickly, exposing the blackness of water and the mystery of dim mud bottom.

Blood root, arbutus and partridge berry begin to show themselves once again. Sphagnum moss, a tenant of acid bogs, is present here in mats of growing, dying and compressed vegetable matter. This adaptable plant is one of the primary constituents of a floating or "quaking bog." Generations of sphagnum, growing and dying year after year, form a great floating carpet of plant life that will support (somewhat tenuously) the weight of a man, and will contribute to the inexorable natural process that converts a pond to dry land. This process would seem to be an unstoppable one, as with each fall of leaves and scattering of windborne dust, the bottom rises, and adventurous marsh plants creep out into the water, claiming for the land ever more space.

At the edge of the beaver pond

At the pond's southern rim, where the water is rendered practically stationary by the beaver dam, a stand of cattails has taken up residence. Perhaps the advance guard arrived in the form of seeds blown into this spot on a chance breeze from a parent stand miles away. Now in early spring they are brown and dried, a few forlorn "punks" or cattail heads protruding into the wind through the rotting ice showing only the promise of future lush growth. Cattails are valuable in many ways. The succulent roots, or tubers, are favored fare of muskrats and men alike, and the dried punks made dandy "smokers," which, when lighted at the kitchen stove, used to give forth a pungent smoke that at least seemed to keep the mosquitoes of August nights at bay!

Bird activity is increasing, by the hour it seems. Near the ledge, the chickadees fill the air with their high, excited chatter, bouncing like animated, black and white ping pong balls through the thickest branches and briars. God help the errant owl that might be uncovered by this exuberant crew! White breasted nuthatches abound. They are the commoner of the two northern nuthatches here of which the red breasted species is by far the rarer in winter.

A striking contrast against the hickory bark, one of these gray-blue and white little creatures hitches its busy way down the trunk, carefully inspecting each crack, prying up flakes and slabs of bark for hidden prey, all the while talking to itself in a low, regular chip-chip-chip.

Tame little birds, nuthatches, like chickadees, seem to know that they have little to fear from people and will allow a closer than usual approach. Unlike crows and gamebirds that sense men may have designs on their hides and give us a wide berth, these little sprites are often the only birds one will see on a winter hike, suddenly materializing out of the trees. Curiosity? Looking for a handout? An offer of peanuts or suet held in the hand is an offer seldom spurned by the chickadees in even the remotest tracts.

Sounds are beginning to gather again—dispelling the long silences of winter. Above, against the racing, low-hung clouds, a mixed flock of grackles and redwings passes over, all talking to each other as they fly strong and swift, driven by a common gathering instinct. Their voices fade as they disappear to the northeast. A robin, too. For a moment it looked like a dried orange perched on a branch until it moved. Am I so unused to seeing color among the greys and white of winter that, for a fleeting instant, my mind's eye did not say robin — it said orange? The bird, a male as told by his richer, pumpkin-colored breast feathers and slate back, perches beside the shriveled carcass of an apple, one of the few left on the old tree, and has obviously been taking nourishment from its dessicated hide in the absence of more usual robin fare. Robins, common as they are, have long fascinated me. As a child, I'd watch them running about our suburban lawn and wondered what that sunset-hued breast felt like, what it was made of. It looked so rich and plush, like brushed velvet, so warm and living! And then their annual peregrinations, the mystery of migration; few have not taken heart at the sight and sound of the "first robin." *Turdus migratorius* epitomizes in the minds of most people the seasonal movements of birds, much as the bear brings to mind (erroneously) the term "hibernation." Actually, robins are not the fly-by-nights they seem to be, deserting us at the first sign of snow flake; though there is a withdrawal southward in winter, many a hardy individual will be seen braving the cold and pecking out a living as far north as New Hampshire and Maine, especially along the somewhat warmer coastal areas. But, by and large, *migratorius* is an apt and descriptive name, for most robins endowed with even a modicum of sense prefer to wait out winter in the sunbelt.

I've now reached the far perimeter of the fifty acre woods, a gentle, rocky knob that gives a comanding view of the surrounding mountains; there to the northwest stands Stinson Mountain, still frosted white over its entire summit and ridge and sporting a barely visible abandoned fire tower at the very peak. Though its hiking trails are frequented in the summer months, Stinson has both the mystery and the threat that all big mountains have. Last fall, the skeleton of a man was found in its rugged untraveled forests by a hunter. The man, the pilot of a small plane that had crashed, had vanished the year before!

To the west lies Tenney Mountain, a "cultivated" peak. Broad, cleared areas of ski trails, looking like great runnels of spilled white paint, crawl down its forested flank. Thousands of local ski buffs and down-staters descend (or perhaps ascend would be a better word) upon the region, creating millions of dollars worth of jobs and revenue in the pursuit of their frost-bitten pleasure.

Sweeping the mountains with field glasses, one can see, over a wide range, the retreat of the snow. The valleys and lower elevations are going first — sizeable sections are bare and the lacework of brown is creeping steadily upward to the mountain top. Here, in this surprisingly early spring, the last shadowed scrap of dirty blue-white will fade away by the month's end.

The knoll is far enough away from civilization that not even the sound of the interstate highway penetrates the pines. This is a spot where in times past I have seen the tracks of bobcat, fisher and fox, all creatures that avoid contact with men and their towns as much as possible. And here today, I make an exciting discovery; a bear "tappen." Looking something like a giant owl pellet, a tappen is long mass of indigestible items such as pine needles, tree bark and hair that the drowsy bear eats in fall before retiring. This intestinal plug is believed to retard digestive action for the winter's duration, and allow the bear to sleep deeply on, with no untoward stomach rumbles to disturb him. The size and length of the plug can give some idea as to how long the bear remained inactive, though this factor is often determined by the severity of outside conditions. My find today indicates that at least one resident bear had probably denned up in the vicinity, as the animals usually eliminate the tappen soon after waking. Resisting a foolish impulse to look warily back over my shoulder, I move on, this time back in the direction of the road that skirts the northern boundary of the woods. Low spots throughout this area are rapidly filling with water — tiny cataracts take advantage of every natural gully and rush down slope and hillside with musical abandon. Coming soon in these boggy, rock strewn glades is the time of the ferns. These graceful tropic-looking plants lend an appearance of lushness and greenery to areas that might otherwise be bereft of it. The many varieties, grape, royal, interrupted, rattlesnake, ostrich (the best edible one), Boston,

Moose

and bracket ferns, can all be found in the spring and summer woodlands of eastern North America. It is hard to imagine, when gazing across the stark winter landscape, that plants as ethereal as these (the slender, beautiful maiden hair ferns in particular) can spring anew from beneath such cold hostile surroundings!

There is so much to be told about ferns! A large and diverse group, they are most plentiful in the tropics and grow everywhere from bogs and moist meadows to cliffs and the trunks of trees. The characteristic "fiddle head" shape of the emerging leaves is familiar to all who explore the woodlands of spring and a salad of those of the ostrich fern is an epicure's delight. The fiddleheads and stems of the common cinnamon fern are covered with fine silky hairs, silvery when the plants are young and attaining later the rusty color that give the plant its common name. These hairs, which seem to be a mere incidental decoration of the plant, figure importantly in the life of another living thing that would, at first guess, have no interest in a fern at all: a hummingbird. These tiny avian buzzbombs have discovered in the course of their evolution that the downy hairs of the cinnamon fern make an ideal building material for their elfin nests. In addition to helping hold together the lichens and tiny twigs which form the bulk of a hummer's nest, the cinnamon hairs provide a soft cradle for the eggs and young which will soon make their appearance in the cleverly concealed bower. In May, the parent hummers can often be seen hovering about stands of cinnamon fern, expertly stripping the fuzz from the stems and packing it off to their home sites—a truly remarkable "harvest" in nature. Today though, the ferneries are silent, flattened leaves in chill water. The living ferns of yesteryear are but prostrate brown stalks, lying along the ground like splayed fingers emanating from the dormant parent root.

So many things are there to be seen by the eyes conditioned to look for the small elements in life. In early spring woods, a flicker of motion may betray the first flight of a mourning cloak butterfly. A species known and loved world-wide, the Camberwell Beauty (its English name) is the first of the *Lepidoptera* to venture forth, virtually on the heels of winter's retreat. Walking up a slope, with the sun just beginning to emerge from behind the overcast, I encounter another butterfly, one that I was certainly not expecting to see: the eastern tailed blue. Here is a true elusive nymph of the woods. The blues are among our smallest butterflies; bright cerulean on the inner wings, greyish silver on their outer sides, they seem to vanish then reappear as they filt erratically through wooded glens and over lawns, bent on unknown butterfly journeys. They appear early in the year, but today's low temperatures and sparse flower opportunities would seem to make this particular blue a short-lived one. On it dances, just ahead of me, vainly seeking out some favored flower species, flicking like a lilliputian semaphore, then darting off to the right and vanishing among the emerging colors of the spring woods.

The story of the tailed blue butterfly involves another strange animal partnership. Among insect larvae, those of these butterflies hold a special attraction for ants, who seek them out and attend them, stroking them with their antennae until the fat grubs exude a sweetish fluid which the ants devour with apparent relish. Such a wealth of organized activity is carried out at a level so far beneath our notice!

The swell of gas fumes in the woods—blue smoke among the trees. Hydrocarbons do not compliment hyla crucifer! Today, with an effect similar to that of a gunshot in church, I encounter trailbikers creating muddy ruts in the woods. I've long wondered what draws the adherents of off-road vehicles to forest travel. Despite the lengthy and often near-convincing arguments by the two-cylinder fraternity about "the lure of the trail", "enjoying nature" and "comradeship", anyone who walks the woods cannot help but voice the angry opinion that mechanical conveyances belong on the roads which, after all, were constructed for them. But once again, as in the arguments that surround hunting, paradoxes abound.

Who does more damage to the forest—the man in Paramus, New Jersey who demands wood for his weekend constructions, thus contributing directly to the decimation of the ancient forests; or the trailbiker who (generally) sticks to his established trails, but shatters the esthetic beauty of the woods, and inhibits the natural intercourse and travels of its wild denizens?

To me a moot point, as with a wave and mud-splattered grin (they are affable, everyday people) the three bikers roar and fishtail by, sending up plumes of grime and wet leaves in their passing. What effect do they have upon the land? The signature of a single trail bike bouncing and spattering along a pine-needle trail is one that will not endure long, but magnify that by a thousand—indeed a million fold, and you have the problem as the environmentalist sees it. The nation-wide epidemic of mechanized man, divorced from his beginnings and bent on conquering the alien world of nature, an offensive movement on the defensive, intolerant of criticism and restriction; this is the problem before us.

And here to, the other side of the spectrum; an intrepid camper in the late winter woods. The man who I met on the trail was out on a foray similar to my own, one of discovery which he shared with me as I did mine with him. I accepted an invitation to visit his camp, revolving around a bright blue scrap of tent perched upon hillock of rock and pine, surrounded by windfall timber and snow. With this hardy human creature I could well ally my own senses and beliefs. Here he was, in the woods at a time of year that few others would brave, below freezing at night, windy unsettled days with the ever present threat of precipitation in either crystal or liquid form! His tent, a sturdy canvas affair with a metal "pipe thimble" snugly tucked into the flaps, seemed admirably suited to comfortable camping in the worst weather winter might throw at him. When I asked him about his heating setup, he at once launched into an enthusiastic description of his approach to the problem of keeping warm and preparing food in a blowing, swaying tent in the middle of the winter woods: "First I took a five-gallon oil can, cleaned it out thoroughly and cut a square door in its side and a circular hole to accommodate a five-inch stove pipe in the top. Then I made the thimble, which insulates the hot pipe from the tent canvas, by riveting two eighteen-inch squares of aluminum together. Flare all four sides, so as to be able to securely tuck the tent flaps in all around it. Two holes drilled through it will enable you to tie it to the tent flaps to hold it in place and another five-inch hole will admit the stove pipe. I set the stove on two bricks to keep it off the can-

vas tent floor, run the five-inch pipe up and elbow it out through the thimble and light up, using dead wood as fuel!" He went on to say that a damper in the pipe was a must for maintaining a low fire at night, and that he felt it better to direct the pipe out at an angle, through the flaps, rather than straight through the roof due to the danger of falling sparks, and snow entering, and eventually rusting out the unit.

What ingenuity is involved in winter camping!

Knowing that my camping friend, like most who travel and set up lodging in the winter woods, had a singular desire not to encounter too many others of his kind, I cut my visit short and bade him and his breeze-whipped tent site farewell. Moving off through the trees, I glance back and see him off on another mission in the opposite direction perhaps to gather firewood or to investigate some natural feature of the land that intrigued him prior to my interruption. A modern man in search of peace, in a world sadly lacking it.

"Summer is icumenin" as the ancient English verse has it, and walking today, one can feel the loosening up, the preparations going on all about. Activity and emerging color in the rich sheltering carpet of leaf and loam just below and eluding the grasp of the senses, lends an air of expectancy much like that experienced by a gardener scanning his fertilized and newly seeded plot. Look at the ground pine on either side of the trail. It is a club moss. The members of its family (*Lycopodium*) are evident all winter long and can be seen (in rather sickly yellowish form) where the ground remains bare. In snow country, the club mosses are just about the first touch of green one will see on an early spring hike. Ground pine is also called Christmas pine due to its frequent use as a yuletide decoration. It spreads by means of runners (I once pulled one up that extended nine feet over and under the leaves), and each plant, in summer, sports a cluster of long yellowish cones that make this common woodland plant easy to identify. Another species, the shining club moss, is very prevalent in acid-soiled bogs and low spots, often forming extensive, plush carpets over rocks, roots and almost any stationary object. Frequently confused with and called sphagnum moss (which is an unrelated peat moss), shining club moss is often used as a natural packing for live plants, and does moderately well in terraria provided light levels are kept low and moisture levels high.

At a high point on the gradual slope that forms the foot of the nearest mountain, I've stumbled upon what must be one of the sources of the rushing brook that feeds the pond further down. Whether this freshet dries up in summer is rather hard to tell, but although there are some tender mosses and ferns in evidence along its course, the water, for the most part, seems to run over leaves and rocks that must be dry and brittle come August. The water appears out of bare, muddy ground between rocks that may, because of a high mineral content and the tracks of at least three deer that are imprinted there, serve as a "lick" for forest wildlife, especially the herbivores. Springs such as this one often carry the dissolved minerals of the subterranean rocks and earth through which they pass on their way to the surface, thus making these life-sustaining elements available to creatures that didn't get them through their regular diets. One of the most productive deer hunting spots will be in the vicinity of such a lick, and hunters with less than admirable scruples will often "salt" a likely wet spot through the year in hopes of luring a trophy buck to it, come November.

I love to follow water courses. Water, as we all know, is such an essential element that wherever it is, there will also be an abundance of life. Though this brooklet is still icy cold and dives now and then, out of sight beneath the snow banks, it is like an oasis in the surrounding and apparently empty woods. Early midges dance in scant numbers over its tumultuous surface; here and there a solitary water strider braces the cold in the quieter pools, and spottings of snow fleas cover other puddles. As I descend, the stream is joined by others and grows in volume. It turns to the east in a deep gully protected by a grove of pine and hemlock and enters winter again. Here the snow is still two feet deep and walking is difficult. Without the aid of snowshoes, I sink deep and flounder ungracefully. Unlike the south facing slope I just left, which was warm with growing sun and nearly devoid of snow cover, this little valley is still deep in winter's clutch. Though sunlight slants in over the treetops and touches the slope I just left and can still see, the very air here is chilled while colors are muted cold greens and blues and nothing moves save an occasional wind- stirred pine bough. Winter leaves us only reluctantly, holding on in these small, fiercely defended outposts.

But out into the sunlight again. Now our water course is a true brook; it emerges from beneath the snow of the gulley with its rocks clothed in rich green fontinalis. In one larger pool, a young trout (a native trout, not a farm-fed stocker I'm sure) senses my approach, and darts under an overhang. Though they "slow down" in the dead of winter, trout relish the cold, oxygen-charged water. This one has probably been hunting caddis worms for a month already.

As I descend in elevation, the air temperature is rising albeit almost imperceptibly. Here and there the furtive new shoots of skunk cabbage and Indian poke, two plants that resemble each other and are often confused, are just beginning to show among the wet leaves adjoining the water's flow. A few early snow-drops, another late winter blooming flower, garland the brookside, and here, where any given square yard of ground has the appearance of a skillfully set up terrarium, I remember an espcially favored spring project of childhood: the survey terrarium. For those so inclined toward investigation of the little things that share our world, I give the particulars herewith. All that is needed is a one- or two-gallon aquarium (or other suitable container), a piece of glass as a lid to completely cover it, a trowel and a pad and pencil. That's it. The procedure is as follows: line the bottom of the tank with sand or gravel, and take it into the woods or fields. With the trowel, dig up a section of turf or forest floor the same size as the tank and place it in the aquarium. This should be set in an indoor location receiving about two hours of direct sunlight daily, watered, and covered tightly with the glass. Then wait. Soon plants will begin to grow, midges, mites and spiders emerge and a virtual army of hidden spores and seeds will stir and send forth their dormant life. All of these happenings can be noted in the ledger so that at the end of a two-month period a reasonably accurate "head count" of the lilliputian population of a square foot of forest or pasture can be had, and the census-taker will find himself much enriched by the experience!

Down off the mountain. The brook is now that familiar stream that will empty its cargo of oxygen, food and silt into the placid pond of the town park. Winter's end is close by now—the exuberent flocks of evening grosbeak and redwings grow noisier by the day. Even at this moment, the sun seems to have gained strength, lending truth to the old axiom "Rain at dawn, sun at noon". Grackles are busy in the fallen leaves, scratching and tossing the debris about in their search for food. Accustomed as they are to the presence of people, they fly off long before I approach close enough to do them any possible harm. Scrapes of carbon paper blown off through the branches!

A pileated woodpecker has flown across my path. A big crow-sized bird with a chinese red crest, it, like its near-extinct relative, the trumpet-voiced ivory bill, is a symbol of untouched wilderness. Last year, we were pleasantly shocked to see one of these black and white-patterned woodpeckers swoop up onto the trunk of a dying maple that stood on our road and begin to noisily hack out a hole in search of its larvae fare. It stayed pretty much the entire morning, despite passing cars and pedestrians, thus displaying an adaptability not suspected in one of nature's greatest lovers of solitude and dead-tree jungles, a true prince of the unmanaged and unmanicured woodlands. To the pileated, a dead tree is the embodiment of earthly beauty!

Just to the south of the fifty-acre woods lay fields that, though shorn of their crop of hay each summer, are not put to any other real use. At their edge, where scraps of snow cover linger the longest, are places of alder and poplar; wet, swampy areas that attract that enigmatic, long billed misfit among shore birds, the woodcock. These ethereal creatures appear in spring often before the snow has disappeared, and their early May courtship displays are something to see and hear. Take a chill late April night, a spring peeper night, and visit the woodcock fields. Listen hard, for that song of the "timberdoodle" is a sound that can so easily lose itself in the spring cacophony of nature's other amorous callers, the rush of wind through yet bare branches, or water over rocks. Soon though, if you're fortunate and your ears are sharp, you'll pick it out: Preent! Preent! Like a forcefully delivered, high-pitched "Bronx Cheer", the odd noise reaches out of the gathering dusk, and appears to come from the edge of the dirt road ahead. A careful approach now — and there he sits on his stage, a place of flattened dead grass, short tail fanned, bill pointed earthward and strutting like a dimunitive rooster! Preent! He's off! Circling around the field, impossible to see against the black forest background, but gaining altitude in one large upward spiral. And now his exuberant flight song drops from the heights, like falling blossoms given voice, and directed at his paramour who sits in attentive admiration in some secluded swale below. The liquid, rolling, chippering notes continues as he circles on high, often visible as a tiny bat-like speck against the day's last light. Then, his pent up energy exhausted, he begins his long swift, zig-zagging descent, appearing like a shadow on the same spot, to repeat the whole performance again.

Such is the woodcock in spring, its only time of noticeable activity. It is a challenge to the sportsman, and joy to the epicure (even the intestines and their partially digested wormy contents are flambed in cognac and avidly eaten as "trail"!). The big-eyed "shore-birds" are not well named. Favoring aspen and alder bogs and wet woodlands, the woodcock rarely turns up near any sort of shore save that of a forest stream or pond. What would an early spring walk be without happening upon the borings and long-toed tracks of the birds in wet leaves and mud greenly studded with emerging skunk cabbage? Probing deeply with their long flexible bills, the timberdoodles seek out their mud-worm prey and extract them from the ground so skillfully that rarely is one even broken in the process. Later in the season, when the callow chicks are afoot behind the parent bird, the solemn little families will be seen probing energetically, the long-beaked kids keeping in touch with mom with scarcely discernable, thin wiry peeps.

My day in the woods, like a thoroughly enjoyed meal of exotic delights, is coming to an end. Everywhere here, where the lay of the land slopes on a south-facing pitch toward the road, the ground is covered with the dark shiny leaves and bright red berries of partridge pea, causing me to think once again of these sturdy winter-resistant birds. Attempts to introduce ringnecked pheasant and quail into this north country have ended in failure; these southern cousins need to be pampered by Mother Nature with areas of bare ground and abundant

Woodcock

wild foods. The deep snows, meager pickings and overall spartan existence to which a mountain winter subjects the grouse is beyond the survival abilities of the others. The winter table is spread thinly here in January, but with the return of spring insects and adventurous buds, *Bonasa umbellus* takes heart. The greening woods will soon echo with the muffled, rapid-fire wingbeats of the courting cocks as they cut the air with stout pinions while perched upon a favored moss-covered log.

Ah, the vernal sun again; it has burned and driven back the overcast and now slants through the trees, throwing the long shadows across the hill road. It seems almost as though every living thing that has emerged to date is bent on squeezing the last minute of activity from the declining day. Bars of sunlight dance and glitter with short-lived midges, a happy crowd of sable crows passes overhead lost in a contest of cries and loud conversation; the flash of a woodpecker (a hairy from its size) among the trees; the distant vibrant clacking of a chorus of wood frogs in some hidden forest puddle.

All of these sights and sounds give voice to spring. The growing things underfoot, the moss carpet, lichens and liverworts, give a hint of the riot of color soon to come. Spring is promise, one can hear it in the voices of children in the park today, returning, like bright colored birds to a place denied them by the will and force of nature through long cold months. Look at the sky this evening. No longer is it that pale blue-white canopy of January, but rather, with the softening of the wind and return of heat, it seems to change, enriching itself and deepening its tone as if preparing a seed bed for the bulky white clouds that are a part of summer's sky.

Down the road toward home. A road, especially a rural one, is a place where often the best of both man's and nature's worlds can be observed. This road, these woodlands and all they hold, have taken me through winter and into spring. As though to place a signature at the end of winter, or perhaps a preface to the coming time of plenty, a fox emerges from the brush at the road's edge and stands looking at me. Caught in the brazen sun, he is beautiful in his coat of burnt orange, and he holds a limp chipmunk in his jaws, one he has just caught, perhaps along one of the stone walls. For the moment we

Red fox with prey

stand and appraise each other; he has no intention of giving up his prize, but is yet hesitant to cross the open road. Finally he does go, flowing across, sun and shadow flashing across his body, the chipmunk swaying, and vanishes into the alders on the other side, the white tip of his plume my last sight of him.

From far down the road a tiny black dot grows and becomes a running dog, ears aflap and legs, very short legs, a moving blur. It is my own dog, an adventurous, but ill-equipped dachsund with a stout heart and a penchant for digging up moles in the lawn. He has one now, a fat plush-grey common mole that is still alive and twists in a futile effort to escape, waving its broad pink digging claws helplessly. It makes no sound. Endeavoring to take it away from him and free it I approach the dog but seeing my intent he skirts around me at a distance, for, like his wild cousin the fox, he has no desire to give up his hard-won catch. He brought it to me to show his digging prowess, not to provide me with nourishment, so off he runs back toward the house to dispose of the mole in his own way.

Walking further down the road puts my own house in view over the trees, the smoke of the last of the winter's cord wood a soft grey plume at the chimney top, drifting off in a gentle wind. This is my own sanctuary, my provider of civilized creature-comforts, barrier against cold and wet and the shimmering heat of summer. But for me, one of a growing army of people divorced from, yet longing for, the "wilderness experience" a home is something that is acquired on different terms than, say, a burrow is by a woodchuck, or the rock ledge den is by the bobcat. They occupy and maintain theirs by the fact of their presence; we maintain ownership of ours through a complex relationship with the tax collector and the bank! The chuck defends his home with his life — when he dies, another moves in. Our system is no longer so elementary, our homes can be taken from our custody by fellow creatures that may have no physical contact with us at all. Such are the ways of man and nature.

In the big maple behind the barn, a lone grackle rasps out the very beginnings of a courtship song that will soon swell from a hundred thousand avian throats. It is the eternal song of spring, the paean

Gray squirrel

of the renewal of life. The barn's steeply pitched roof supports a circular, shrinking patch of snow, which yields to the late sun in a curtain of drops raining from the roof's edge. With the lowering of the sun behind the dark trees in the west, a deepening cold sets in and one by one, the droplets slow down and stop. Tonight ice will re-form on still water and quiet will prevail in the winter woods.

TRAVEL IN
THE WINTER WOODS

Most pleasure tourers in the winter woods will confine their travels to the space of one day and not attempt, unless forced, to spend the night outdoors. Trail blazing and wilderness survival is best left, at the moment, to those well-steeped in experience. But day trips into even a few acres of January woodland can do much to develop a "wilderness sense" and build the stamina and skills necessary for extended, overnight hikes.

One requisite for winter hiking is adequate clothing. Should you become temporarily lost, the powerful reality of cold will become immediately apparent if you have not used some foresight in the selection of warm, insulating clothes. Since freedom of body movement is also essential, garments should not be tight, constricting or more bulky or weighty than necessary. At the least, the following should be worn on a mid-winter walk: long underwear, outer shirt of wool or jersey, soft weave or wool pants, heavy socks, either a parka or down ski jacket, and, last but certainly not least, a wool stocking or ski cap, big enough to be pulled down to fully cover the vulnerable ears. Mittens or gloves are a must, heavy waterproofed boots, a pocket compass, matches in a waterproof case, sunglasses and a topographical map of the region to be traveled (if extensive), will about round it out, except for food and water and something, such as a knapsack, in which to carry what you are not wearing.

On a day-long trek, the careful selection of foods will obviously not be of paramount importance, except that the addition of quick energy-producing items such as pure, milk chocolate bars, would not be a bad idea. The question of libation is another matter. The ideal liquid addition to the grub pack is a thermos of hot coffee or chocolate which will provide both warmth and stimulation and is much preferable to cold water after a long climb through heavy powder. If water is preferred, it's best to carry your own in a two-quarter canteen, running it luke-warm from the tap, and carrying it in the shelter of the pack to retard its freezing. Cold water should be drunk in slow sips, not gulped, especially when one is overheated after exertion. In my own experience I've found it best not to trust natural running water sources within five miles of population centers unless you are familiar with the area and know the water to be pure. In general, my own rule of thumb is that a mountain brook, found close to its source among the rock strata and supporting a healthy population of aquatic plants and pure-water fish, such as dace and trout, may be considered a safe enough risk in winter, spring and early summer. But beware; conditions will vary tremendously throughout the range of the winter woods. A clear brook running through Connecticut farm country may carry a sizeable load of pesticide residue, coliform and typhoid bacteria and Lord knows what else, while tumbling stream in the Great Gulf wilderness of northern New Hampshire, cloudy with spring runoff silt, may be better than the most thoroughly treated tap water! Play it safe and pack in your own supply.

The smoking of an occasional cigarette or well-tamped pipefull might seem to be an innocuous and relaxing indulgence but, aside from obvious health reasons, on a long hike the smoker should refrain at least for the time being. Nicotine is a vasoconstrictor, hampering the flow of blood through the blood vessels making them more vulnerable to cold and frost bite. When it occurs, frost bite, or hypothermia, is not best treated by rubbing the affected part with snow. The usual field procedure is to place the affected member against a warm part of the body, such as in the armpit (if it is the hand), or wrap it well in a scarf or other such covering, and get to medical help as soon as possible. Do *not* place a frostbitten body member in hot water or under hot compresses.

Alcohol is unwelcome on a January field trip since it may give the body a sense of allaying cold while in fact acting as an anesthetic rather than a stimulant, thus lowering resistance to and awareness of the dangers of freezing tissues. Some parts of the body, such as the ears or the toes, are likely to freeze without one's being aware of it, particularly if their owner has dulled his senses to any degree beforehand. The use of any substance which alters one's view of his surroundings and their potential dangers certainly has no place on the winter trail. Especially if one is traveling alone in territory that is too extensive to be readily searched in one day, carelessness resulting from a drug or liquor may help secure a top spot on the casualty list.

For the novice, it's always best to go out with a companion, no matter how harmless or domesticated the tract may be. A mossy ravine, easily explored in summer, may be a snowdrift-filled death trap for the unwary snowshoer in winter. It's much more fun to share experiences with a fellow traveler and, should you get into some inexplicable jam, it's a lot better to have help in seconds, rather than in hours. When out on snowshoes alone, the best possible advice would be to take along ski poles or a stout walking stick. In the absence of the helping hands of a trail mate, the leverage of a walking stick or ski pole may be all that stands between you and a wait for searchers.

Righting oneself in deep soft snow after a full length tumble on snowshoes can be a problem. Unless there is a handy sapling nearby to provide leverage, trying to regain one's feet in such a situation is roughly similar to attempting to stand up on the surface of the water while swimming. Rather than thrusting the arms down to find purchase on solid ground, it is better to draw yourself into a crouching position, pulling the snowshoes underneath until you feel balanced and then stand up. A good walking stick will provide great assistance under such conditions and will also be of use in testing snow-depth, the dependability of ice and defense against the over aggressive dogs one might well meet on a hike.

On the subject of snowshoes, the question of size and length is

dependent upon snow depth and texture. Packed snow with a crust will not require a large shoe whereas deep soft powder will. Among the varieties of snowshoes, the Michigan, or Long shoe is about the best all-around pattern. Bear Paws are used mostly for short trips, or by trappers where a great deal of maneuverability is required, though they are not suitable for large or heavy persons. The Beavertail is a wide shoe suitable for touring, and is a good general purpose shoe. Proficiency on snowshoes takes practice. All the book-reading on the subject will not make you an expert your first time out. Be wary, careful; learn to avoid spots that may contain submerged tree roots and lengths of barbed wire lying just underneath the snow. Strive to develop a sense of balance when descending steep slopes and you'll soon be a veteran snow traveler. Recognizing these potential hazards takes time and often painful experience, but, as in most acquired skills, there is no other way.

Direction-finding on a hike of moderate length and in familiar territory will, of course, present no problems. One can orient himself from natural features of the terrain or a distant horizon. In unfamiliar country, it's wise to carry a pocket compass, noting the direction of north and your own path of travel at the start. A topographic map of the region is an added asset. You can orient the map by aligning its north with the northerly direction in the field, and so marking it.

The old adage that "moss grows on the north side of the trees" is not too reliable, particularly in dense conifer forests and north facing slopes where it will grow everywhere. The position of the sun, if its position is visible, especially in winter when it never rises high above the southern horizon, is perhaps the best natural method for establishing general direction. Again, even a rough advance knowledge of the location of roads, streams, or a town will be an aid in the prevention of getting lost in all but the most extensive forests.

For the best results in the observation of the life in the winter woods, one should simply follow the rules of all good field technique:

Be methodical—form a plan of action beforehand, deciding just what you'd like to accomplish (photography, for instance) and take along the necessary materials. Binoculars are almost always useful.

Hooded mergansers

Be leisurely—allow yourself enough time to cover the area with patience and with frequent stops to scan your surroundings.

Be quiet—though dry, crackling leaves will not present a problem in snow country, dead twigs and brush will. Nothing pops louder than a snapped branch on a still winter day! Windfalls and underbrush should be negotiated with care if wildlife observation is the object. On animal-spotting trips in familiar terrain, traveling alone is most productive.

Be alert—proceed with all your senses attuned to your surroundings, scanning both the near and far distances, watching for any movement.

Be inquisitive—investigate what you see and hear—you'll most surely be rewarded for the effort.

Small mammals can be attracted to the observer simply by remaining still in a secluded spot and using one of the commercially-made predator calls. I have found these particularly effective with mink, weasel and fox. Whatever small birds are along the line of travel can best be attracted by the fine art of 'pishing'. This "homemade" predator call is, in my opinion, the most effective way of calling up passerine birds, having a longer range and more urgent sound to it than the old squeaking-on-the-back-of-the-hand method. To produce the sound, make believe your are loudly attempting to quiet noisy kids. Cup your hands around your mouth, and have at it. Pssh-pssh-pssh! Repeat the call in several directions. If there are any avian eardrums within range of your efforts, your should get results, and fast.

By and large, the responsible tourer in the winter woods is the one who best observes the same rules and courtesies practiced when entering the home of another fellow human. The woods, summer or winter, is home to many more forms of life than any of us will ever realize and our responsibility to them and to their habitat is no less than that which we feel towards any other.

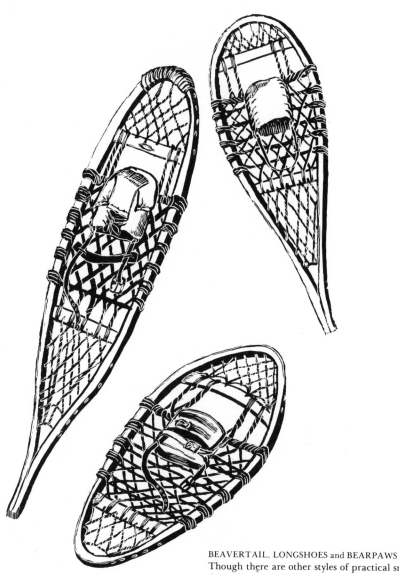

BEAVERTAIL, LONGSHOES and BEARPAWS
Though there are other styles of practical snowshoes, these three are the main types for use in the winter woods. The beavertail (upper) is an old Indian design, good for soft powder. The long, or Maine snowshoe (center) is the one used by the author. Long and narrow, it will support a person of heavy weight, yet does not require a spraddle legged stance, and is good for long hikes. Bearpaws are smaller, rounded shoes, best for persons of lighter weight and scoring higher points in maneuverability. They are favored by trappers and for use near camp or homestead.

SPECIES APPENDIX

The following pages present, with greater detail than in the text illustrations, some of the more common plants and animals which might be found in the winter woods.

Because the animals are more wary and less active than in other seasons, none but the most fortunate winter hikers will actually observe more than a few on any single venture into the woods. Depending on the nature of the terrain and its location, one may also come upon many forms of life other than those covered in this basic collection. Discovery of the unexpected and the unusual is one of the greatest delights for the naturalist in the field. The habitat of the winter woods is just as likely to provide such surprises as any other.

BAGWORM COCOON
If the hanging cocoon of the bagworm is examined in winter, it will usually be empty, though at times the dormant, unhatched eggs of the moth will be found.

CECROPIA MOTH COCOON
(*Samia cecropia*)
The cocoon of this big, nocturnal summer moth is found on winter branches. The species ranges from the Atlantic coast to the Great Plains.

BLACK FLY
(*Simuliidae*)
Scourge of the north woods, this hard biting, tiny fly appears in spring, usually near dense woodlands and water courses.

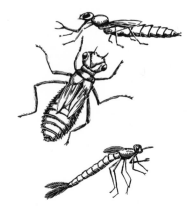

CADDIS WORMS
(*Trichoptera*)
These tiny larvae, lugging about their "houses" of weeds, sticks or pebbles, may be observed in winter in quiet shallow water. The adult flies are regarded as ancestors of butterflies and moths.

DRAGON and DAMSEL FLY NYMPHS
These small, but fearsome-looking insects are often observed, in a torpid state, in winter brooks. The dragon fly larva (center) is an active predator of young fish.

Assorted WINTER GALLS
Pictures are some of the galls commonly seen on bare winter twigs and branches. They are the homes of eggs and larvae, and most are produced by sawflies and related insects.

MOURNING CLOAK BUTTERFLY
(*Anglais antiopa*)
A butterfly of the leafless, early spring woods, this species hibernates in adult form.

MAYFLY NYMPH
(*Ephemera varia*)
These curious-looking nymphs are found living under stones in brooks. The adult flies are a favorite food of fish.

PRAYING MANTIS COCOONS
(*Mantidae*)
Mantis cocoons are a common sight on dried winter stalks. The egg mass of the introduced mantis (*M. religiosa*) is larger than that of the native species. (lower fig.)

SNOWFLEA
(*Collembola*)
Many species of these primitive insects appear in great numbers on the late winter snows across North America.

EASTERN NEWT
(*Triturus*)
The common "salamander" of lakes and
ponds. Most country and suburban
children have seen and kept this species.

AMERICAN BEECH
(*Fagus grandifolia*)
Smooth blue-grey bark, and pale, buff colored leaves clinging to the tree identify the
beech in winter. To seventy to 100 feet.

SPOTTED SALAMANDER
(*Ambystoma*)
A large, conspicuous salamander seen in
ponds in early spring.

NORTHERN WHITE CEDAR
(*Thuja occidentalis*)
Northern white cedars grow in limestone
and boggy soils. Forty to fifty feet.

SPRING PEEPER
(*Hyla cricifer*)
This tiny frog is a true harbinger of spring.
Its bell-like calls are known to all who live
near ponds and bogs. About one inch.

BALSAM FIR
(*Abies balsamea*)
This tree grows to its highest in wet soils near lakes and streams. To sixty feet.

EASTERN HEMLOCK
(*Tsuga canadensis*)
The bark on the mature tree is scaly, and a dark purplish-brown. To sixty-five feet.

TREE FUNGI
Many species of tree or bracket fungi may be seen in the winter woods. Shown are three representative species: the birch leuzite (top); the oyster pleurotus (center); and the multi-zoned polystictus (bottom).

EASTERN WHITE PINE
(*Pinus strobus*)
The common pine of the east, it is also the largest conifer there, reaching heights of 100 feet.

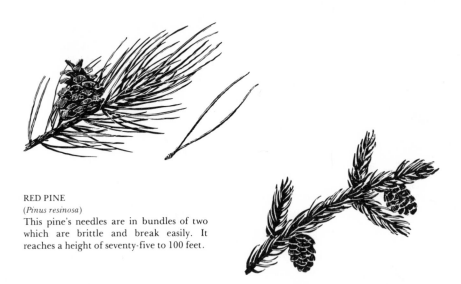

RED PINE
(*Pinus resinosa*)
This pine's needles are in bundles of two which are brittle and break easily. It reaches a height of seventy-five to 100 feet.

BLACK SPRUCE
(*Picia mariana*)
This tree grows best in boggy situations and reaches a height of forty feet.

TAMARACK
(*Larix larcina*)
The tall, graceful tamarack is a tree of the boreal forests. It averages forty to eighty feet.

BROOK TROUT and BLACK NOSED DACE
(*Salvelinus* and *Rhinicthys*)

These two species are among the several that might be seen in winter brooks and ponds. The brook trout (above) is the one most likely to be observed, preferring cold water at all times of the year. The mottled sculpin or "Miller's Thumb," a curious, big headed fish, may also be found hiding under brook stones in the coldest water.

BEAVER
(*Castor canadensis*)

Active all winter, but not seen by the hiker as they remain in the lodges or below the ice.

BLACK BEAR
(*Ursus americanus*)

This is the largest land carnivore in the east. Black bears sleep through the severest winter weather, but do *not* hibernate!

WHITE-TAILED DEER
(*Dama virginianus*)
This species, as well as the mule and black-tailed deer of the west, remains active all winter, though at a low level of activity to conserve energy and fat reserves. Their yards are quite vulnerable to disturbance in the form of snow-machines and dogs.

EASTERN FLYING SQUIRREL
(*Glaucomys volans* & *sabrinus*)
Though not uncommon throughout most of its range, the flying squirrel is seldom seen, even by experienced travelers of the winter woods. Active throughout most of the winter, these little squirrels breed in March or early April in the north country.

FISHER
(*Martes pennanti*)
This big furbearer of the north woods is more at home in the trees than on the ground. The smaller, but closely related marten is also a swift arboreal (tree dwelling) animal that preys upon squirrels and any other small creatures it can capture. Both are active all winter.

RED FOX
(*Vulpes fulva*)
The Beau Brummel of the woods and fields of North America, the wiley, beautiful red fox has managed to hold its own within the presence of man. The fox is active in winter and is often seen in the woods.

DEER MOUSE
(*Peromyscus maniculatus*)
This secretive, little-known (to the average person) mouse is one of the most abundant small mammals of the woodlands. Somewhat larger than the obnoxious house mouse, the deer mouse is primarily a nocturnal animal.

SNOWSHOE HARE
(*Lepus americanus*)
In summer, the snowshoe hare is a large brown rabbit. In winter it is pure white, save for its eyes and ear-tips. The meandering big-footed tracks are distinctive.

PORCUPINE
(*Erethizon dorsatum*)
This big, lumbering rodent is active throughout the winter season, living then on the inner bark of such trees as white pine, hemlock and maple. Look for them high in the treetops in winter. Porcupine tracks show the wide pigeon-toed stance of this rodent. Animal tracks, especially in winter snow, seldom look as clear-cut and sharp as they appear in diagram form, hence one must learn to recognize them through general shape and pattern.

SHORT-TAILED SHREW
(*Blarina brevicauda*)
Though several species of shrews may be encountered in the winter woods, this one is the most common. They are active all winter and their distinctive, delicate tracks identify them immediately.

RACCOON
(*Procyon lotor*)
The raccoon is a highly adaptable, resourceful animal, that is about for much of the winter, though sleeping through the worst of weather. They are found in all manner of habitats, from suburban yards to the deep woods. Usually, the hind tracks overlap those of the forefeet.

RED SQUIRREL
(*Tamiasciurus hudsonicus*)
These active, noisy squirrels are one of the most noticeable denizens of the winter woods, spending more time on the ground than the larger grey squirrel. They are known to tunnel for long distances through snow drifts. The red squirrel does not bury nuts singly, but establishes food caches.

STRIPED SKUNK
(*Mephitis mephitis*)
The adaptable skunk is found everywhere from the deep forests to farms and suburban yards. Skunks apend part of the winter season denned up, but wander forth during the mating season in February and March.

SHORT-TAILED WEASEL
(*Mustela erminea*)
An alert predator, active all winter. The weasel captures shrews, mice, etc., throughout snow season.

WOODCHUCK
(*Marmota monax*)

The woodchuck, or ground hog, is not a true winter animal. Disappearing below ground in October or November, depending upon latitude, it does not reappear in the fields until about April or May. Woodchucks are true hibernators.

BLUE JAY
(*Cyanocitta cristata*)

The bright blue and white jay is familiar to all throughout winter country. It is found in coniferous-oak woods and frequents feeders in winter. The grey and white Canada jay, or camp robber, is a bird of the north woods and is exceedingly bold and fearless.

MEADOW VOLE
(*Microtus pennsylvanicus*)

All animal life on earth, including man, is ultimately dependent upon the plant kingdom for food, and the vole is very near the bottom of the food chain. It feeds upon green plants and is in turn preyed upon by an army of predators. Active all winter.

120/THE WINTER WOODS

CHICKADEE
(*Parus atricapillus*)
Almost everyone knows what a chickadee looks like. Industrious and aggressive, these little birds are found throughout forested areas, and frequent suburban bird feeders as well.

CROWS
(*Corvus brachyrhynchos*)
These birds need no description. While they favor suburban and rural areas, crows are so adatable they are found virtually everywhere.

BROWN CREEPER
(*Certhia familiaris*)
The brown creeper is a secretive little sprite of the woodlands. This bird always works its way spiraly up a tree trunk, then flies to the base of another tree.

EVENING GROSBEAK
(*Hesperiphona vespertina*)

These stocky gold, black and white birds
with the big white bills are a common sight
throughout most of the winter woods, fre-
quenting bird feeders where they prefer
sunflower seeds. The related pine
grosbeak, a larger, rose and grey colored
bird, is also a visitor from the northern
forests.

NUTHATCH
(*Sitta carolinensis*)

The white-breasted nuthatch is a familiar
denizen of the winter woods, working its
way up and down tree trunks in search of
insect food. The smaller red-breasted
nuthatch (*S. canadensis*) is another winter
species.

RUFFED GROUSE
(*Bonasa umbellus*)

This relative of the domestic chicken remains active all winter, holing up under
the snow during extreme cold spells. Their
meandering "chicken tracks" are a familiar
sight in good grouse habitat.

HOUSE SPARROW — top
(*Passer domesticus*)
STARLING
(*Sturnus vulgaris*)
Both of these species, introduced from
Europe, are found virtually everywhere in
more urban and farm areas. In the north
country, they are more or less restricted to
the towns, where there is a ready food sup-
ply. Both of these birds visit winter feeders.

BARRED OWL
(*Strix varia*)
This is the "eight-hooter" of the southern
woodlands, but it is also found in good
numbers in the snowbound forests of the
north. The smaller, yellow-eyed, ear-
tufted screech owl is also active through the
winter season.

DOWNY WOODPECKER
(*Dendrocopos pubescens*)

This woodpecker, along with the larger, but similar hairy woodpecker, is found in the trees in winter. Their single, high-pitched "peek!" calls are distinctive.

PILEATED WOODPECKER
(*Dryocopus pileatus*)

The pileated woodpecker is a big, crow-sized, black and white bird that is only occasionally seen in the winter woods. A relative of the nearly extinct ivory-billed woodpecker, this bird chops great holes in the sides of dead trees in search of insect larvae and grubs.